From the moment of conception

until delivery nine months later, the human being is more susceptible to his environment than he will ever be in his life again.

In this fascinating and important book Ashley Montagu details a sound and practical program of prenatal care. He explains the importance of nutrition to the developing infant; just what happens to the embryo when the mother is emotionally overwrought; how drugs can be harmful to the unborn child. Some of the medical information the author provides reads like an old wives' tale . . . in particular the proven fact that a child can actually learn before birth!

Thoroughly documented with clinical studies and the results of medical research across the world, *Life Before Birth* offers invaluable counsel to all couples of childbearing age.

SIGNET Titles of Special Interest

Ashley Montagu

LIFE BEFORE BIRTH

FOREWORD BY
Alan F. Guttmacher, M.D.

A SIGNET BOOK from
NEW AMERICAN LIBRARY
TIMES MIRROR

To Charlotte and Gayle Aiken

SIGNET TRADEMARK REG. U.S. PAT. OFF. AND FOREIGN COUNTRIES
REGISTERED TRADEMARK—MARCA REGISTRADA
HECHO EN CHICAGO, U.S.A.

SIGNET, SIGNET CLASSICS, SIGNETTE, MENTOR AND PLUME BOOKS
are published by The New American Library, Inc.,
1301 Avenue of the Americas, New York, New York 10019

FIRST PRINTING, JUNE, 1965

PRINTED IN THE UNITED STATES OF AMERICA

FOREWORD

DR. MONTAGU has written a fascinating book full of pertinent material, material especially pertinent to couples planning a family, but also to couples already pregnant. However, my long life in obstetrics causes me to issue a word of caution to the reader. Don't be frightened by some of the things you will read. Remember that mother nature is on your side and has been for the one million years man or manlike creatures have peopled the earth. If not, I would not be here to write these words and you would not be here to read them. Remember above all that the vast majority of pregnancies turn out magnificently well and most newborn babies are perfect specimens. Dr. Montagu is simply trying to reduce to zero the very small percentage of pregnancies which go badly and the small percentage of babies which are not born perfect—truly a commendable goal.

In the main I agree with his advice, but don't panic if circumstances or even your own weak will causes one of Dr. Montagu's admonitions to go unheeded. Perhaps you crave a cigarette so badly that you forget or you simply disobey. The chances for a few cigarettes to cause any obstetrical difficulty such as early labor are too remote to brood over. Then, too, if Dr. Montagu's conclusions are at variance with your own doctor's advice for you, do what your doctor says—to him you are an individual and your welfare and that of your baby is just as important as it is to you. I am sure your doctor agrees with Dr.

Montagu that unnecessary X rays during pregnancy should be avoided, but at the same time you may have a problem which he feels necessitates X ray. Let your doctor judge. In the same way your doctor, after carefully weighing the pros and cons in your individual case, may decide to induce labor by injecting Pitocin into an arm vein. He has specific reasons for you, which outweigh the generalizations of any author, even so well informed an author as Dr. Montagu.

Therefore, my friends, read *Life Before Birth* with interest and attention, and if you are able to follow its advice, your chances for what would probably be a very happy experience anyway are increased—even if you follow most of its advice, this is still true. You are lucky. Birth has never been as safe for mother or baby as it is this year, and with Dr. Montagu's advice it will become even safer.

Alan F. Guttmacher, M.D.

PREFACE

THIS BOOK has been written for everyone interested in human beings, but especially for prospective mothers. Life begins, not at birth, but at conception, and what happens in the interval between conception and birth is very much more important for our subsequent growth and development than we have, until recently, realized.

Our original setting is in our mother's womb, and the experiences we undergo in it are fundamentally important. In that setting our foundations are laid. What happens to us afterward is closely related to that time we spent in our mother's womb, to what happened to us in it. Something of this was dramatically brought to public attention by the thalidomide tragedy of 1961–1962.

As one who has spent the greater part of his academic life in the medical school and the hospital, and at the same time in the field of anthropology, as teacher and researcher, I have for many years been concerned with the growth and development of human beings long before they become children and adults. My real interest has been in discovering what happens to human beings in the womb that may affect their subsequent development, both physically and behaviorally, outside the womb, to the last days of their lives. In February 1962 I published a technical work on the subject.* It is a large and expensive book. It is not going to reach all those who ought to be acquainted with the facts it contains, especially mothers—

* _Prenatal Influences_. Springfield, Ill.: Charles C Thomas, 1962.

potential and those in process. Hence the present volume.

This book assumes no technical knowledge whatsoever. In the task of translating my technical work into its present form I have depended upon the skill and enthusiasm of Suzanne Fremon. I am very grateful to her. To Katherine Stechmann I am especially indebted for her excellent critical and editorial work on the manuscript.

A.M.

CONTENTS

I—EVENTS OF GREATER MOMENT

THE PRINCIPAL PURPOSE of this book is to show a prospective mother the various ways in which she can influence the physical and emotional development of her child before he is born. This is not, however, a "how-to" book, nor a set of rules, nor yet a course in physiology, although it contains some of the elements of all of these. It is an attempt to explain and evaluate for that part of the population most intimately concerned—the women who are having the children—the enormous progress that has been made in recent years, and that continues to be made all day every day, in the understanding and care of children before they are born.

This book contains no guarantees. In the basic matters of human life, there are no guarantees. It does contain facts, figures, conjectures, conclusions, and recommendations. These last could, if they were assiduously followed, change the ratio of illness to health in the entire human species.

Every woman wants her baby to be born perfectly healthy, with sturdy bones, good vision, normal organs throughout his body, and the promise of growing into a healthy, intelligent child. As a matter of statistical fact, if a woman lives a comfortable, relaxed, healthful life herself all during her pregnancy, the chances are overwhelming that her child will be born in superlative condition anyway. But, thanks to the work of many thousands of scientists all over the world, we now know that she can push his chances up even higher by doing certain things, avoiding others, and taking certain precautions.

11

What these are, how they work, and why—so far as this is known—constitutes the substance of this book.

Men, too, as fathers, may find these pages interesting and useful; a man is more fully a father, and he is certainly more helpful and companionable to his pregnant wife, if he knows approximately as much as she about the complex processes that make a healthy baby.

The basic fact is simple: life begins, not at birth, but at conception.

This means that a developing child is alive not only in the sense that he is composed of living tissues, but also in the sense that from the moment of his conception, *things happen to him*. Furthermore, when things happen to him, even though he may be only two weeks old, and he looks more like a creature from another world than a human being, and his birth date is eight and a half months in the future, *he reacts*. In spite of his newness and his appearance, he is a living, striving human being from the very beginning.

His mother can help him most during these earliest days by comprehending this fact and by controlling as closely as she can the things that happen to him. This is a relatively new concept in medicine: the vital importance of prenatal influences. And the proof is newer still.

Even a doctor who tries to keep up with developments in medicine may not be aware of all the work being done. "Prenatal influences," he muses. "Interesting field . . . but do we know much about it?"

We know a lot about it. First of all, we know this: During the prenatal period—that is, the nine months between conception and birth—a human being is more susceptible to his environment than he will ever be again in his life. What happens to him then can help him in his development, or it can hinder him; it can promote his growth or it can interrupt it.

Secondly, we know this: The events that take place before his birth, and his reactions to them, will influence him for the rest of his life. When he is twenty or forty or sixty there will still be aspects of his body or his mind or his emotions that will be traceable directly to the influences that acted on him before he was born.

The food his mother eats, her general physical condition, her immediate state of health, how old she is, how tired she becomes—these are some of the factors that are part of the baby's environment. These, plus numerous others, separately and together, will largely determine whether or not he will be born healthy and normal.

All these influences have been studied, and are still being studied, in laboratories and hospitals and clinics in many parts of the world. This work is far from finished; in fact, it will probably never be finished. But enough has been done by now so that when we gather it together and look at it all at once, we see that it forms an arrow, pointing unmistakably toward a new understanding. It is time, then, to organize this material and evaluate it, so that mothers and fathers of the new generation of children may learn the lessons that are implied in it and the children themselves may benefit from it.

A few general recommendations to mothers are already emerging. The most important of these are listed here, but they and others are discussed more fully in the appropriate sections of the book.

1. Good nutrition is probably the most important single factor in the development of a child before he is born. Therefore, a pregnant woman—indeed, any woman who will ever be pregnant in the future—should be certain that her nutrition is not only adequate, but good. This is particularly true during hot weather, when most people tend to skimp on proteins. A developing child, especially during the first twelve weeks of his life, when his major organs are being formed, needs proteins more urgently than he will ever need them again, and his mother is his only source of supply.

2. A pregnant woman should not take any drug whatever during her pregnancy, unless her own life or health depends on it, as in the case of insulin for diabetes. There are several reasons for this, and they are described and discussed in detail in Chapter VI.

3. She should not smoke. The reasons for this are given in Chapter VII.

4. She and her doctor should use anesthetics and analgesics with great care. During the birth of the child, a regional anesthetic is better for the baby than a general anesthetic. This is because a general anesthetic, like the substances in cigarette smoke, cuts down the amount of

oxygen in the bloodstream. A further explanation of this phenomenon can be found on pages 99–116.

5. A pregnant woman should not have treatment with X rays unless it is absolutely essential for her health, and she should be extremely careful with X rays for diagnostic purposes, even in such small exposures as a dentist would give her for a routine examination of her teeth. (See pages 204–220.)

6. Sometimes protective vaccination and other immunizations are recommended, because some diseases and infections can travel from the mother to the child. These are discussed in detail in Chapter X.

7. A pregnant woman should be careful not to become over-tired. For this reason she should probably not work at a fatiguing job outside the home later than the fourth or fifth month of pregnancy. (See Chapter XI.)

8. She should try to remain as unperturbed as possible; her emotional condition is important to her baby. The interesting studies on this subject are outlined in Chapter XI.

9. She should do everything possible to avoid having a premature child. If this threatens, she should be particularly careful to follow her doctor's instructions. Although these children are now magnificently cared for, a child gets off to a better start if he is born at term.

10. Special situations, such as those that go along with surgery during pregnancy, or the birth of twins or triplets, or birth by Cesarean section, are handled best when they are planned and prepared for well ahead of time. (Chapter XII.)

11. Very young mothers and mothers over thirty-five should take particular care with their nutrition, their rest, and their general health. The general subject of the mother's age is discussed in Chapter V.

12. Since a child who is only two or three *weeks* from conception can respond to his environment, every woman who is in her childbearing years should follow these same recommendations at all times if she wants to ensure her future children the best possible chance of health. In other words, even if she thinks she is not pregnant, she should be careful; she may find out later that she has been pregnant all along.

How do we arrive at all this? Some of it we recognize

as common knowledge: we have known for generations that nutrition, for example, is of primary importance, and the continuing studies of the nutrition of pregnant women are among the clearest and most convincing of all the work done in this field. But emotions? Smoking? Anesthetics? Do we really *know* that these things are part of a child's immediate surroundings before he is born?

The answer is Yes. We know this because of the work of thousands of scientists who have spent years of their lives investigating these very things, asking these questions and many others besides, and slowly, meticulously, finding the answers.

This round dozen recommendations, and many more suggestions that will be mentioned as we go along, have been deduced from the enormous body of published reports that describe each investigation in minute detail. Some of this work is done as a by-product, so to speak, of a big-city hospital clinic, where a single doctor is suddenly struck by the unexpected appearance of some disorder among newborn children. He wonders; he ponders; he begins to investigate. If he is on the right track, he may suddenly see, in a flash of insight, what he has been searching for.

Something like this happened in 1941 in Australia, when Dr. N. M. Gregg, an ophthalmologist, was called upon to examine an unusually large number of children who had been born with cataracts. Dr. Gregg's flash of insight made a connection in his mind—although such a connection had never before been made—between the outbreak of congenital cataracts and a recent epidemic of German measles. He looked into the histories of the mothers of these children, wondering whether he would find a connection between the two outbreaks. He did. All the mothers had had German measles.

There were only thirteen children involved in this study, but they were enough. Dr. Gregg reported the thirteen cases in a medical journal and included his discovery that their mothers had had German measles, his conclusion that the two items were related, and his conjecture that the German measles in the mothers had caused the cataracts in the children.

Thus, quietly and without fuss, are historic discoveries sometimes made. Immediately, Dr. Gregg received answering confirmation from all over the world: other doctors had begun to suspect the same connection, and they were coming, only a little later than Dr. Gregg, to the same conclusion.

Some investigations are planned in advance and carried out according to a schedule, with the idea of proving or disproving a connection that is already suspected. Such a study was made on a large scale of a group of undernourished women in a prenatal clinic of a large hospital. At the beginning of pregnancy, some of the women were given supplementary diets, to bring their nutrition up to a high level. Other women were given some extra food, but not so much as the first group. When the children were born, they were examined and compared. The children of the women who had been eating well were in much better condition than the children of the mothers on the poorer diets.

Still other important conclusions are drawn from a gathering together of many small reports on the same subject. A private physician may see only a few instances of the various rare conditions of human beings, but when he observes carefully, describes completely what he sees, and then publishes his report—and almost all doctors do this—and when numbers of other physicians add their reports, the trickles coming in from all over the world over the years swell into a flood of information. Someone then comes along who wants to know what other people have seen of this condition; he searches, he finds, he pulls everything together, he writes *his* report, and the medical profession knows something it didn't know before.

Other information lies hidden away in file drawers until the probing questions and brilliant imagination of a curious man bring it to light. For example, the records of the Columbus State School, in Ohio, over a period of thirty-six years, were searched by an enterprising pair of medical investigators in an effort to determine whether there was any correlation between season of birth and mental retardation. The results of this fascinating and revealing study are given in detail on pages 198–199.

In recent years a new force has entered the field of medical research. This is made up of the private foundations and associations that support, with more lavish funds than most individual investigators can command, studies in specific fields. The National Foundation is a good example of this. Having solved the problem of the prevention of poliomyelitis, this foundation has now turned its attention to the support of research designed to show the world how best to prevent congenital abnormalities. The Association for the Aid of Crippled Children is another group of this kind.

As we glance across the surface of the work being done these days, it suddenly seems that we have shifted our gaze from healthy, normal children to children who are ill or in some way unfortunate. Is this necessary? Is it not possible for scientists to show us how to produce healthier children, without concentrating their attention solely on children who are not healthy? Dr. Gregg, in Australia, seemed to be ignoring all the thousands of children there who were born perfectly normal, to focus his attention on the thirteen children who were not. If we are primarily interested in healthy children, perhaps we would do better to concentrate on them.

But *was* Dr. Gregg ignoring the healthy children? And the doctors who conducted the study on nutrition—were they concentrating only on the sickly babies? Of course not. The nub of the matter here lies in a comparison between two groups of children, and in every study this comparison is made. Here we have a group of sturdy, normal children; there, on the other hand, is a group that is suffering from some specific ailment. Is there some one event, or factor, or influence, that is both common to the ailing children and lacking in the healthy children? It is only in this way that investigators can put their fingers on a particular spot and say with confidence, "This is the villain!"

Or to study the problem from the other direction: here we have a group of women, all of whom had Asian flu during pregnancy. And on the other hand, here is a group in which none had it. How do their children compare? Is there something that is common to most of the children in the first group that is lacking in the children of the second? If there is, maybe the Asian flu was responsible.

This sounds simpler than it is, in practice, because perhaps it wasn't the flu, after all, but something that goes along with flu, that caused the ailment in the children. A drug that is commonly given to treat flu, for example, might have been responsible. And then one wonders: what about their backgrounds? Perhaps the women who had the flu were chiefly women whose resistance was low and whose general health was poor; this, in turn, might have been due to poor nutrition. If this is true, then it might be more accurate to say that the children of the mothers with flu were not so big and healthy as the other children because of the differences in nutrition. What a tangle of cause and effect!

Another tangle lies in the intertwining threads in every human being that we label "heredity" and "environment." When we consider the behavior of an individual, we sometimes try to comb out these threads, and when investigators are studying a specific item in the complexity of human nature, they must proceed as though they could reach in, pick up a single thread and hold it to the light, and say, for instance, "This is clearly the result of environment."

To a certain degree this is justifiable. The little packages of chemicals in our bodies, known as genes, do indeed carry the potentialities for various characteristics from parent to child, and from child to grandchild, and so on down the generations. Similarly, a surrounding environment does exist, and it is composed of particular factors that can be itemized.

However, as soon as we begin to consider the development of a child from a single cell to a complete and functioning human being, the lines separating the two areas begin to blur. The genes alone cannot produce a human being; there can be no development at all without an environment to provide nourishment, warmth, and protection. And yet the same environment acts on different people in different ways. Fraternal, or nonidentical, twins, for example, who had had exactly the same environment for nine months, are usually quite different from each other; this is clearly the effect of the differences in their inheritance.

Furthermore, complicating an already complicated situation, genes provide an environment, as it were, for other genes, and to this extent we can even say that an environment is inherited. In addition, each of us is born into a unique and particular environment. Also, the circumstances surrounding a human being act and react on each other; and they act and react on his genes, developing them, subduing them, occasionally even changing them. The individual human being is the total expression of all these complex and constantly interacting forces.

It is not easy to study such a creature.

Fortunately for individual mothers and children, and for the human race in general, it is not so difficult to act on the conclusions of the studies as it is to conduct them in the first place. The most important step to take in ensuring good health and normal development for a child before he is born is the first step: to recognize that it is both possible and important.

One hundred and fifty years ago, long before our scientists had even begun to suspect the importance of life before birth, a poet, Samuel Taylor Coleridge, wrote, commenting on a passage in Sir Thomas Browne: "Yes,—the history of man for the nine months preceding his birth would, probably, be far more interesting, and contain events of greater moment, than all the three score and ten years that follow it."

Coleridge was right, and the rest of us are just now catching up with him.

II—THE GIANT STEPS

WHEN A HUMAN BEING begins his life, he is smaller than the period at the end of this sentence. He is a single cell, barely visible to the naked eye, a tiny drop of fluid made up mostly of nonliving material, with an even smaller body called the nucleus—containing the only truly living material—in the center.

When he is born, thirty-eight weeks later, this human being is about 20 inches long, he weighs about 7 pounds, and he is made up of 200 billion cells. Furthermore, the cells have not only multiplied during these weeks; they have also specialized. The newborn baby's body has become an enormously complicated mechanism, composed of hundreds of different kinds of cells.

Never again in his life, in so brief a period, will this human being grow so rapidly or so much, or develop in so many directions, as he does between conception and birth. It is this rate of growth and complexity of development that give his mother the unique opportunity for influencing him. Although we commonly think of infancy and childhood as the periods in a person's life when he is permanently molded by what happens to him, the period immediately preceding that—the prenatal period—is even more important, and for similar reasons. Just as his behavior and attitudes are formed during infancy and childhood, so during the weeks of prenatal life he is being formed physically. His skeleton develops from nothing at all to the elaborate structure of bones that will support him all his life. His organs come into being and develop into the machinery that will take care of his breathing, his circulation, his nourishment, the elimination of waste, and reproduction. The muscles appear and spread them-

selves about his body in the intricate network that will
govern his physical movements. Most marvelous of all
these marvels, his brain and nervous system will set up
the complicated arrangements that will make it possible
for him to see, smell, feel, and even think. And every
one of these systems is dependent, for its successful development, on its immediate surroundings during the prenatal period.

This is surely the broadest field in all the life of a human being for activity and influence. It is, indeed, the
whole man. Later in his life, if he says that a new love
or a new job has changed him, he will be speaking figura-
tively, although he may not altogether realize it. When we
say, here, that a particular influence can change him before he is born, we mean it literally. It can affect the
length of the bones in his legs, or the successful formation of
his liver, or the proper construction of the muscles around
his heart, or the quality of his brain. In short, it can affect his basic equipment.

Furthermore, it can do this in an unexpectedly brief
time. The entire prenatal period is short, indeed, but the
period of major influence is shorter still. It is the first
third of pregnancy that is most important in the child's
development. This is not to say, of course, that the remaining weeks are not also important; nothing that
happens to a child before he is born is unimportant.
However, during the first twelve weeks, the major organs
and basic tissues are being laid down and developed, and
once their development is finished, it is difficult, if not
impossible, to affect them in any fundamental way. Thus,
the first twelve or thirteen weeks of pregnancy are the ones
during which a mother has the greatest opportunity to influence the growth and development of her child.

During this critical period, the development of the human body exhibits the most perfect timing and the most
elaborate correlation that we ever display in our entire
lives. The building and launching of a satellite, involving
thousands of people and hundreds of electronic devices,
is not nearly so complex an operation as the building and
launching of a human being. His development proceeds
in an orderly manner and at a regular rate, with specific
changes occurring at specific times. Every organ and every
tissue—in fact, every cell—has its own timetable for

coming into existence, for developing, for taking its place in the machinery of the body, and for beginning to carry out its functions. And every small timetable is meshed with every other timetable. The whole process is so orderly, in fact, that embryologists have been able to draw up a schedule that accounts for all parts of the human body and shows how each part fits into the whole: a countdown, in effect, that starts with the instant of conception and goes on until birth.

This countdown also shows us how it is possible to affect the development of a child from the outside in specific ways by specific agents. For example: The thalidomide episode of the summer of 1962 brings into bold relief two important facts, only one of which has received sufficient public attention. The first is that the drug thalidomide, when taken during the first three months of pregnancy, can seriously harm the development of some children. The other, and perhaps equally important, fact is that this drug, although potentially damaging under adverse conditions, did not harm the children of numerous women who took it during other periods of their pregnancy. In other words, the effect of even such a drug as this is limited both as to kind and as to time. Thalidomide evidently strikes most severely at the centers that govern the growth of the bones of the arms and the legs. When thalidomide is taken into the mother's body at exactly the period during which these centers are being formed, it will, in many cases, prevent their development. The critical phase of toxic thalidomide action has been found to be between the twenty-seventh and fortieth days after conception. At any other time during the entire pregnancy, these centers cannot be affected by this drug. Once they are formed, they are formed for good, and nothing can change them after that.

Thus the critical period in the formation of any single structure or function of the body is brief. It may last only a few hours, or a few days—and even in the case of the most elaborate systems, not more than two or three weeks. And it is during its specific critical period that any structure or function is most susceptible to its environment.

Life begins when a single sperm cell penetrates and fertilizes a single egg cell: this action produces the single living cell with which every one of us starts life. Growth begins when this cell divides into two. Each of these cells then divides into two, making four. The four then divide again, then the eight, the sixteen, the thirty-two, and so on. About forty-four such cell divisions are necessary to build the human body up to the size of a baby when he is born.

But he is no simple cluster of cells when he is born; the cells that make up his body are of many different kinds: bone, skin, muscle, nerve, glands, hair, blood, fat, and numerous other tissues with specialized functions. Each of these cells appears quite different under a microscope from each of the others, and together they combine to form entire systems, each a miracle of organization, and related to other equally complex systems. This activity of cells is known as differentiation.

Differentiation begins the sixth day after conception. Until then, the cells have been dividing, slowly, within the cell wall of the original fertilized egg, until there is a cluster of about a hundred cells, not much bigger in volume than the first cell. By the sixth day, the cells in this cluster begin to separate into two layers of cells, and to change their nature. The outer layer will eventually be part of the placenta—the organ through which the child and his mother conduct most of their business. The inner layer will before long begin to give rise to the body of the child. This is the first of the many differentiations that will end thirty-seven weeks later in a finished human body.

At about the seventh day, the fertilized egg (sometimes called an *ovum,* from the Latin word for "egg") sinks into the wall of the uterus. From the outer layer of cells now grow many fine, fingerlike projections called *villi,* which penetrate even more deeply into the wall of the uterus, down to the muscle, where they come to rest in the blood spaces of the mother. These villi can be thought of as roots, reaching out toward the sources of nourishment. During the fourth month of pregnancy, they will combine with part of the lining of the uterus to form the placenta, and it is mainly through this combination of villi and placenta that the mother's body sends the various materials to her child to help him grow and develop.

Her blood cells, carrying on their surfaces molecules of protein, of carbohydrates, of minerals and vitamins, of oxygen and other chemicals, bathe the villi. And just as the roots of a plant are fed through their contact with the earth, so the villi take up the nourishment from the mother's blood. The substances on the surface of her blood are passed through the thin walls of the villi, and from there enter the child's circulation.

The growth of the placenta and the establishment of the fully developed system between mother and child take place gradually over the first three months of pregnancy. At the same time, many other cell divisions and differentiations are going on within the child's body.

First, the inner layer of cells begins to differentiate. Three separate groups of cells eventually emerge from this layer, and they place themselves in such a relation to one another as to form cavities within the whole mass. The chief of these is the amniotic cavity, where the baby will later lie.

By the time the baby—no longer an ovum, but now properly called an embryo—is fifteen or sixteen days old, he is the shape of a round, flat plate. From that, he changes to the shape of a pear. Although he still has a considerable distance to go before he looks like a baby, this is the beginning of his human appearance, because the broad end of the pear-shaped embryo will eventually develop into his head. Down his length a groove appears; then a ridge rises on either side of the groove. The ridges eventually come together all down the length to form a tube. During the next two weeks the upper end of this tube will begin to develop into the child's brain and the lower end into his spinal cord. He is now on his way.

Then growth and development accelerate. A thin wall appears across the body of the cells: this will later be the diaphragm, separating the heart and lungs from the stomach, liver, and spleen. The child begins to double up, because he is growing faster than the cavity in which he is lying, and this change of position helps to form the openings of the gastrointestinal system and the urogenital system. The esophagus, the stomach, and the intestines begin to take shape, and muscle and connective tissue begin to appear around these organs.

By about the beginning of the fourth week of the child's life, he begins to develop his own circulatory system. The villi, the little roots that reach down into the mother's blood, now contain blood vessels of their own. A portion of the inner group of cells narrows down, changes into the umbilical cord, and blood vessels grow within this cord, to link together the growing vessels in the embryo and those in the villi. During the fourth week the baby's heart begins to beat, and from that time on, his blood circulates in ever-increasing volume, carrying oxygen and nourishment deposited in the mother's bloodstream to the child's liver and heart; from there it is pumped to every part of his body, and then returned, now carrying carbon dioxide, nitrogen, and waste products, to the villi, and from there into the mother's bloodstream. At no time do the two bloodstreams mingle; they are entirely independent. Blood cells are normally far too large to pass through the walls of the villi, or later, when it develops, through any part of the placenta. By the end of the seventh week, the child's circulation is functioning.

This is a short period—from the close of the fourth week to the close of the seventh week—for the development of such a complex system as this, to serve the individual for the rest of his life. And the development of the circulatory system is not all that happens during this three-week period: the stomach and intestines are forming at the same time; the entire nervous system of the body —the brain, the spinal cord, and the nerves—is developing; the respiratory system, with the passages from the mouth and nose as well as the lungs, is coming into being; the reproductive organs, the glands, the liver, the kidneys, the bladder, all make their appearance during these twenty-one days.

The child's arms and legs begin to grow now, too, first as buds, then as finlike shapes, then recognizably as arms and legs. A platelike formation appears at the end of each arm and leg—the future hands and feet. Rays begin to grow from them; these are the fingers and toes.

The eyes and ears are defined, too, at the beginning of the three-week period, and as the days go by, the child's face appears. The nostrils are formed. The eyelid folds, the top of the ear, and the little hillock that will become the nose are all visible by the seventh week. The size of the head increases greatly.

Before the end of the seventh week, the child's head elevates, his neck can be seen, and his trunk straightens. He is still far from the finished product he will be thirty weeks later, but he is clearly human, and now he has most of the organs that he will need for the rest of his life.

By the end of eight weeks, he graduates again: properly speaking, he is no longer an embryo but a fetus. This word comes from a Latin word meaning "offspring," and the change in labels at this stage symbolizes the child's growth from a rudimentary being to a recognizable human creature.

During the next month, the child's third month of life, his nails begin to form, the rest of his bones begin to develop, and it is possible to recognize his sex from his external genitals. His face is clear, and in unlike twins the differences begin to be recognizable. He is now nearly as complex in his physical make-up as he ever will be.

During the child's fourth month, the formation of the placenta will be completed, and the remainder of his prenatal life can be devoted primarily to growing. That is, he has developed his major systems by now, and although they will still change to some extent, the chief change from now on will be in size. And for this spurt of growth the child needs the assistance of that remarkable organ, the placenta.

The placenta's primary function is to provide a passage through which the necessary amounts of nourishment and oxygen can reach the child, and through which his waste products can be funneled away. Its importance is immediately obvious: without it, a child could not grow. A healthy, well-developed placenta is as necessary to the proper development of a child as a well-functioning maternal heart or liver.

The placenta forms slowly, more slowly than any other major organ, between the inner wall of the uterus and the outer wall of the embryo. When it is full-grown, almost at term, it is round and flat, about 7 inches in diameter and 1 inch thick, and it weighs about 1 pound. The side of the placenta facing the embryo is smooth, and from it springs the umbilical cord; this, containing large blood vessels, joins the body of the child. The side of the placenta facing the mother is covered with the fine, rootlike villi, containing even finer blood vessels. These roots burrow deep into the wall of the uterus, where the mother's blood is plentiful. The number of these villi is enormous, considering the modest size of the organ from which they spring: that number has been estimated at 770,000. And the total area of their surfaces, through which the molecules of food and oxygen pass, has been estimated at 140 square feet.

The passage of this material is basically simple. The walls of the villi are composed of a series of several layers of cells, relatively thick in the early placenta, but becoming steadily thinner as pregnancy advances. This thinning process occurs because of the increasing size of the child and his growing need for food and oxygen. The placenta is not able to grow in proportion; therefore, it must become more permeable.

The bloodstream of the mother carries the materials her child needs—proteins, minerals, vitamins, and the like— in the form of molecules of various sizes. The smaller molecules, encountering the villi, find their way through the permeable surfaces and in this way enter the child's bloodstream. Larger molecules, too large to penetrate the walls of the villi, are sometimes broken down into smaller, simpler substances and pass through in that form.

The placenta is a remarkably efficient organ. It appears to allow the passage of just enough material to supply the child's needs, no more. Oxygen, for example, passes through the walls of the placenta so slowly that the blood of the child contains no more oxygen than he requires. For this reason it is particularly important to guard against interfering with this flow; there is not much margin for safety in this area.

All gases, being composed of small molecules, are capable of being transmitted from the mother to the child. This is an important fact to remember in an age when many women smoke, and most women live in some degree of industrial smog.

Numerous drugs, too, as we know from newspaper reading as well as from studies made by medical investigators, are capable of crossing into the child's circulatory system. Anesthetics and analgesics pass through easily, and so do some substances that are poisonous. Some chemicals that are needed by the child cross to him easily from his mother: iron, calcium, phosphorus, sugar, and common salt are among them, and all these materials can be stored by him in his system and used when they are needed. Other chemicals must be broken down before they can be transmitted; still others must be in the presence of assisting substances before they can be pushed through. Amino acids, proteins, and fructose pass through the placenta slowly; glucose goes through rapidly. Sometimes the presence of various enzymes will help the proteins and the sugars across the barrier. Fats are not transmitted directly from mother to child, but are absorbed by the placenta, which then changes them rapidly into simpler substances and passes those substances along. This is an example of one of the additional functions of the placenta: it assists in the process of metabolism—that is, the building up and breaking down of food into simpler substances.

Antibodies, those materials that the human body manufactures to combat specific bacteria, can travel from mother to child, and from child to mother. This is frequently a great protection to both. The bacteria themselves must "eat" their way through, breaking down the tissues and then absorbing them. To bacteria that cannot do this, the placenta acts as an impenetrable barrier.

Viruses—at least some of them, and perhaps all—as well as the poisons they produce, are capable of passing from mother to child. Hormones, the chemical substances that are manufactured by various organs of the body and are carried in the bloodstream to other organs where they stimulate activity, can be transmitted through the placenta. In fact, the placenta itself manufactures hormones, which travel from the placenta to several parts of the mother's body.

This hormone-manufacturing function of the placenta emphasizes the fact that this organ, temporary though it is, plays more than one important role in the bodies of both mother and child during pregnancy. In addition to being the principal channel through which the child receives his nourishment and his oxygen, the placenta, through the activity of its hormones, largely controls the general progress of the pregnancy. The hormones maintain the proper tone of the uterus and govern its contractions, thereby determining when the pregnancy will end and labor will begin. It is by means of these hormones that the mother's ordinary twenty-eight-day cycle is supplanted during pregnancy by the thirty-eight-week cycle. In fact, thanks to them, the placenta is the most potent endocrine gland in the body of a pregnant woman.

In some situations, the placenta acts as a barrier. Although our emphasis has thus far been on its qualities as a path between mother and child, sometimes it stands as a guard across that path, preventing the entrance of many large-molecule, harmful materials to the body of the child. This double function constitutes one of the most interesting aspects of the placenta.

All these various functions of the placenta, then, are carried out during the last half of pregnancy. But what of the first four weeks of pregnancy, before the placenta has begun to function at all? And after that, when the placenta is still immature and functioning only partially? There must surely be channels other than the placenta by which materials can be transmitted from mother to child. In fact, we know this to be true from occasional accidents that have happened to children during the first few weeks of their lives.

Experiments with animals have shown us something of the ways in which these accidents may have come about. Several investigators have followed the passage of various materials—proteins, sugars, and even vaccines—from the mother's body into the body of the embryo, and have come to the conclusion that these substances find their way from the mother's bloodstream through the membranes of the embryo in the form of droplets of plasma. These are then absorbed into the outer layer of cells of the embryo, where they wait until the embryo develops a functioning circulatory system of his own. Then they pass into his bloodstream.

Later in pregnancy, the child habitually swallows quantities of the amniotic fluid in which he lies. In this way, various materials, beneficial, harmful, and neutral, can enter the child's body.

In general, although materials in the fluids of the uterus and the amniotic sac may find their way into the body of the child, during the last half of pregnancy it is principally by way of the placenta that the child is nourished. And he needs this special organ; he is growing now.

His basic structure is set, his skeleton is in place, his organs are formed, and many of them are functioning. But he is small. The rest of his prenatal life will be devoted mainly to increasing his size. By the time he is born, he will weigh about 7 pounds.

By the end of the fifth month, he will have grown to about 1¼ pounds, and he will be covered over most of his body with downy hair. There may even be some hairs on his head. Also, although he has been capable of muscular movement long before this, he is now able to move so powerfully that his mother begins to feel his movements for the first time.

His proportions improve by the end of the sixth month. His skin is reddish, and because it has been growing faster than the tissue underneath, it is wrinkled. In fact, his appearance now foretells his appearance many years from now, when he is an old man, and his skin will again be looser than the underlying tissue. He now weighs approximately 2¼ pounds.

By the end of the eighth month, fat is being deposited around the child's body, and he has put on an additional 2 pounds in weight. His skin is somewhat less wrinkled, and it has begun to thicken. His eyelids are no longer fused, and his eyebrows and the hair on his head are visible.

At the end of the ninth month, the redness of his skin has faded to pink, and his fingernails and toenails are the perfect miniatures that delight his parents. He is plumper than he has been, although not nearly so plump as he will soon become. He weighs 7 pounds. His legs, like those of an ape, are shorter than his arms. All his organs are now prepared for postnatal functioning. In short, he is ready to begin an independent existence.

When we contemplate the growth and development of a child before birth, two general recommendations for prospective mothers emerge quite naturally from what we see. The first, stemming from the awe with which we properly regard the marvelous machinery that constitutes the human body, is: Leave the machine alone as it develops, and by the time it is built it will be functioning smoothly. Specifically, don't tinker with it; don't sprinkle sand in it or allow anyone else to do so; don't drop a monkey wrench into the mechanism and break it. In short, leave it alone. This is the negative recommendation.

The positive recommendation is: Give the machine the best possible fuel. This fuel is the food and oxygen that the mother sends to her child, chiefly by way of her own bloodstream.

The two together combine to give the developing child the best possible life before birth: nourishing food and oxygen, and protection from harm. The rest of this book will be concerned with specific aspects of this subject: how do we know that nutrition is important? what constitutes good nutrition? what are the possible dangers that a mother must guard her child against? what can she do? how do we know?

By knowing these things, and acting on them, a mother can give her child what he needs most. If he has this much from her, he will be able to take care of the rest of the process himself.

III—THE BEST PROVISION: GOOD NUTRITION

A MOTHER'S NUTRITION is the most important single environmental influence in the life of her unborn child, and it is by means of the food she eats that a mother can have the most profound and lasting effect on her child's development.

This hardly comes as a surprise to an observant person. A healthy woman who has eaten properly all her life nearly always bears a sturdy child who radiates good health. In contrast, a woman whose appearance clearly announces "undernourished" frequently has a child who is thin, pale, and sickly. This observation has been borne out scientifically over and over again, in investigations of large numbers of women and children and in carefully controlled studies of laboratory animals. There is now general agreement among physicians that the difference between good health and poor health in a newborn child lies more often than not in the nutrition of the child's mother during her pregnancy.

A word of caution: Nutrition is also closely interwoven with many other factors. Frequently, even when we can see plainly that nutrition is playing a decisive role, we cannot single it out and label it as the particular cause of a particular result. Good nutrition, for example, is generally connected with a relatively high standard of living, and poor nutrition with a low standard of living, but any consideration of standards of living must include many other factors in addition to nutrition. Education plays a part, too. People who suffer from malnutrition are, with a few exceptions, ignorant of the facts of human growth and development. Thus we can say that poor nutrition is often

32

a *result* of other and deeper factors that are buried in the social, economic, and educational conditions of an entire group of people.

But nutrition is also a clear cause, and it is chiefly this aspect that concerns us here. A good diet in a pregnant woman is a direct cause of healthy growth in her child, and a poor diet is a cause of deficient growth, however these diets are achieved. This is demonstrable, as we shall soon see. In addition, nutrition can be an indirect cause, working in any one of several directions. For example, a mother's good nutrition over her whole lifetime results in the superior development of her own body. If she has grown normally from her own prenatal days—and this depends on her mother's diet during pregnancy—her own organs will be splendidly equal to the job of nurturing her child; if not, they may be less able. On the other hand, a young woman whose own background has been poverty-stricken or deprived can, by providing a good diet for herself and her child, compensate to a large extent for her own deficiencies. In the same way, any woman, whatever her background, can counteract by means of a good diet the possible damaging effects of other less controllable influences.

Dr. Theodosius Dobzhansky, the distinguished geneticist, has used the word "groceries" to cover all the raw materials that a human being takes from the world around him and uses for his growth and development. The use of this simple domestic word, particularly when we apply it narrowly, to food alone, brings home to us, as a more exact scientific word might not do, the daily importance of nutrition. Every one of us needs his groceries every day. An unborn child is no exception. In fact, since he grows more rapidly and develops in more complex ways during these nine months than in any other similar period in his life, he needs adequate groceries more acutely now than he ever will again. And he will suffer more now from inadequate groceries than he ever will later.

He has only one source from which he can obtain these: his mother. She, then, takes on some of the functions of the owner and operator of a grocery store. Her child is her only customer, and she is more fortunate than most grocers in being able to find out, if she takes the trouble, what her customer's daily order is likely to be.

She can, in fact, arrange to have the shelves of her store well stocked. She can, if she will, manage her inventory so that all her customer's needs will be promptly satisfied and that he will never go hungry or suffer in his development from malnutrition.

Perhaps the most dramatic and clear-cut example of this kind of stocking up on behalf of the customers occurred in Switzerland several decades ago. The situation was serious: in several cantons of Switzerland, for many years, a particularly large number of the babies had been born cretins. As a result, cretinism was frequent throughout the population, and since it is a condition of marked deficiency, both physical and mental, it presented a serious problem for the families and the communities where it occurred. In most parts of the world, cretinism is very rare; in certain pockets here and there, however—and a part of Switzerland was one—it is more common. The government of Switzerland, impelled by the need to care for many of these unfortunate people as they grew older, took a hand in the situation. Investigations were carried out, and the reason soon emerged: the entire population was suffering from a lack of iodine in their diet. Iodine is essential to the proper development of the thyroid gland; when it is lacking, the thyroid gland fails to function properly, and cretinism is sometimes the final result.

In most parts of the world, iodine comes to us naturally and in sufficient quantities through our food. In this section of Switzerland, however, and in a few other places, the soil contains almost no iodine. This means that the vegetables and grains grown in that soil are also deficient. It means, too, that the cows who eat the grass have no iodine in their milk, and the animals butchered for meat are also lacking in it. There is, in short, no natural way for the people who live in such an area to get enough iodine in their diets for healthy development.

The Swiss government found a solution quickly: all pregnant women were given supplementary iodine. And the result, even to those who prescribed the iodine and knew precisely what the effect would be, seemed magical. In the course of one generation, cretinism all but vanished.

This episode, thrilling in its significance for the individual Swiss citizens who benefited from it, and for the betterment of human beings in general, is surely the clearest and simplest example we have of two important facts: a lack of proper nutrition in a pregnant woman can have serious consequences for her child; and by the simple act of improving her diet where improvement is necessary, she can greatly influence the development of her child toward normal, healthy growth.

No such large-scale and dramatic example of the importance of nutrition is available to us from the records of our own country, partly because we have been lucky enough in the first place not to suffer such wholesale deficiencies that need such heroic measures. But our medical journals are studded with reports of investigations that prove and clarify the connection between a mother's diet and the development of her child before birth.

One of the earliest of these studies was carried out between 1939 and 1941 in the prenatal clinics of the Boston Lying-in Hospital. The investigators studied over 200 pregnant women and their children. They divided the women into three groups: those who had been on excellent diets, those who had been on good diets, and those who had been on poor diets. The result of the study was immediately revealing: almost all the infants born to mothers on excellent diets were in good or excellent condition when they were born. In contrast, all the stillborn children, all but one of the premature children, and almost all the babies who were born with any difficulties at all were born to the women who had been on poor diets.

A relationship was also discovered in this study between the diet of the mothers and their own physical difficulties during pregnancy and the birth of their children. The women who had been living on good diets had significantly fewer complications of various kinds. Toxemias of pregnancy, a term that covers several common disorders of pregnancy, were much more frequent among the women on poor diets than among those on good diets. Also, the women who had been well nourished experienced much smoother deliveries than the poorly nourished women, in spite of the fact that the well-nourished babies were on the average almost three pounds heavier than the others.

An interesting extension of this study was undertaken later by the same investigators: they followed the experiences of fifty-seven of these women through a second pregnancy, again keeping close track of their diets. At the conclusion of this second study, there was no question in the minds of the investigators that the diet of the mother was of primary importance in the condition of newborn children: whenever the diet of a woman improved over her first pregnancy, she gave birth to a child who was in better condition than the first; when her diet deteriorated, her second child was in poorer condition than the first.

These observations are of highest importance, because there is much less room here than in most studies for confusion among various factors that may have a bearing on the result. The comparisons in this study were made between children of the same parents, living in the same households; the principal difference between each pair of children was in the diet of their mother before their birth. It is not often that two human beings can be compared with so little variation in their situations. Any two children, even of the same family, vary somewhat in their inheritance, and the mothers of some of these children may have been living slightly differently during their second pregnancies, but even making allowances for natural small differences, these studies point clearly to nutrition as the most important factor in their differences in health.

Another group of early studies was undertaken in 1942 in Toronto by Dr. J. H. Ebbs. He and his co-workers cared for and studied 210 women during pregnancy, saw them through delivery, and studied both mothers and children after birth. All of these women, clinic patients, were poor. Furthermore, when they began coming to the clinic, all of them were on inadequate diets, seriously deficient in protein, calcium, and iron, with too few calories.

The clinic provided all these patients with supplementary nutrition, but divided the women into two groups and gave different amounts to each group. The first group received some protein, some calcium, and some iron, but the quantities in each case were too small to be considered an excellent diet. For example, although ideally these women should have been eating 85 to 100 grams of protein every day (or 3 to 3 1/2 ounces) they were getting only 55 grams (or 2 ounces).

In contrast, the remaining women received enough supplementary nourishment to bring their daily diet up to an ideal level. They were given 2,500 calories, 1.6 grams of calcium, and 24 milligrams of iron a day.

The clinical story of these women, set out in Figure 1, is revealing, and confirms the crucial role of nutrition

| | | Diet | |
		Poor	Good
Prenatal maternal record	Poor-Bad	36.0%	9.0%
Condition during labor	Poor-Bad	24.0%	3.0%
Duration of first stage of labor	Primipara	20.3 hours	11.1 hours
	Multipara	15.2 hours	9.5 hours
Convalescence	Poor-Bad	11.5%	3.5%
Record of babies during first two weeks	Poor-Bad	14.0%	0.0%

Illnesses of Babies
  During First Six Months

Frequent colds	21.0%	4.7%
Bronchitis	4.2%	1.5%
Pneumonia	5.5%	1.5%
Rickets	5.5%	0.0%
Tetany	4.2%	0.0%
Dystrophy	7.0%	1.5%
Anemia	25.0%	9.4%
Deaths	2.5%	0.0%

Miscarriages and Infant Deaths

Miscarriages	5.8%	0.0%
Stillbirths	3.3%	0.0%
Deaths:		
Pneumonia	1.7%	0.0%
Prematurity	0.8%	0.0%
Prematures	7.5%	2.2%

FIGURE 1. Comparison of 120 pregnant women on a poor diet with 90 pregnant women of the same socioeconomic status whose diet had been made good (Ebbs).

from conception through birth and well into postnatal life. In general, the findings of the Toronto study agree substantially with those of the earlier study in Boston, although some of the figures and percentages are different. Very few of the women with good diets had any difficulties during pregnancy, during labor, or during convalescence. Many more of the women on poor diets encountered difficulties at all three stages. Among those who were having first babies, labor lasted only half as long among the better-nourished women than the others. None of the babies of the well-fed women had difficulties during their first two weeks of postnatal life; 14 percent of the others had poor records. Only 2 percent of the children of the well-fed women were born prematurely, as against more than 7 percent of the others. The well-fed women had no miscarriages and no stillbirths, and none of their children died; the figures for the others are not nearly so good. During the first six months of the children's lives, their records show a remarkable difference: among the children of the well-nourished women, there was almost no illness, and there were no deaths; among the others there was much more illness, and 2.5 percent of the children died.

One of the most surprising aspects of the study was this: none of the mothers in either group, the well-fed or the poorly fed, showed any signs themselves of malnutrition or deficiency diseases. In other words, the diet of a pregnant woman can be so seriously inadequate that her child is in danger from it, without producing any recognizable symptoms that might act as a warning to her or her doctor. This fact is itself a warning: a prospective mother must analyze her normal diet, with the help of her doctor, and determine whether it is lacking in anything, and if so, in what, and take immediate steps to remedy the deficiency. She cannot count on a warning signal from her own body.

One of Dr. Ebb's colleagues recorded his impressions of another, less measurable, aspect of improved nutrition in the Toronto mothers. He comments that "one of the first evidences of change that was observed in the women when their diets were improved was a change in their mental outlook. Before improvement of their diets they were worried about everything, the whole business was a trouble to them, housework bothered them, they were unhappy. When they came back after three weeks on a good diet, their faces were washed, their hair was brushed; if they could get a new feather they stuck it in their hats. As Sir John Boyd-Orr remarked, 'Don't forget, a woman during the childbearing age who is not interested in her appearance has something wrong physiologically.' These women became definitely interested in their appearance."

An interesting correlation between nutrition and the premature birth of children has been worked out in several countries of Europe. The basis for the work is the hypothesis that prematurity is frequently a result of poor nutrition in the mothers. As we shall see later, this is an important field for study: premature babies run many more risks than babies born at term. The experimental work seems to bear out this hypothesis: in England and Wales, the number of premature children has dropped significantly as a result of giving supplementary diets to pregnant women. Our own prematurity rate is about 10 percent; that is, about 10 percent of all the babies born in the United States are born prematurely. The governments of all the Scandinavian countries have taken advanced measures for the welfare of pregnant women—measures that include good diets, freedom from heavy work, and frequent rest periods. The prematurity rate has been brought down to an average of 5 percent in all these countries, except Finland, which has achieved the remarkable rate of 2.4 percent.

Prematurity is also caused by factors other than nutrition, and several of these can combine to form the pattern that is basically founded in poverty and ignorance. We cannot say, therefore, that supplementary diets for pregnant women would completely wipe out prematurity or any other single effect, but we can say that a woman who lives on a relatively comfortable standard, who learns the facts about nutrition and applies them to her diet, will without question influence strongly the healthy development of her child.

One of the ways we can learn about the influence of a good diet on a pregnant woman and her child is to observe the effects of a poor diet. World War II, unhappily, offered numerous opportunities for this. During the winter of 1942, for example, in the city of Leningrad, while it was being besieged by the German army, the civilians were being bombed and shelled, they were suffering from extreme cold weather coupled with a lack of heat in the houses, and they were forced into a great deal of unaccustomed heavy work because of the disruption of all the normal ways of living. While all this was going on, a doctor in one of the Leningrad hospitals, Dr. A. N. Antonov, was studying the pregnant women and the children who were born during that year. In the early days of the siege, Dr. Antonov found it almost impossible to separate the various factors in the lives of his patients. Inadequate nutrition was clearly one of the problems shared by the entire population, but so were fatigue and anxiety and various infections. As the siege continued, however, he began to observe a phenomenon that plainly indicated the importance of nutrition. This was that although the number of women who became pregnant sharply decreased as the months went by, the average condition of the women who were pregnant, and also the condition of their children, tended to improve. This paradox was finally explained by a combination of two facts.

The first fact was that the siege conditions were so severe for the entire population that large numbers of women ceased entirely to ovulate and to menstruate; this caused the rapid drop in the number of births. The second fact emerged when Dr. Antonov investigated the occupations and family backgrounds of the women who had succeeded in becoming pregnant: they were all employed in jobs where they received extra rations because of the importance of their work. They were doctors, nurses, teachers, or they were wives of military personnel or of workers in particularly important war industries. These women were suffering precisely the same anxieties, fears, cold, and exhaustion as all the other women in Leningrad, but they were eating better than the others. Result: their reproductive systems continued to function, and their children, when they were born, were healthier than the children born earlier in the year to other women on smaller rations.

Later, Dr. Antonov reported on his observations of pregnant women and their children during the three years of siege and famine in Leningrad. It makes unhappy reading, and there would be little point in detailing it except for the fact that it emphasizes, as no peacetime study possibly could do, the importance of the mothers' nutrition for the development of children. For example, during the particularly severe famine period of early 1942, the number of stillborn children was double that of the prewar period. The rate of premature births rose at the same time to the extraordinary peak of 41 percent. Even the children who were born at term showed the effects of the famine: their average weight was 21 ounces less than the prewar average, and the normal three- or four-day period after birth when babies lose weight continued for six days. In addition, they were ill more frequently and more seriously than was common before the siege, and the death rate was much higher.

Other investigations of civilian populations during wartime have borne out these findings from Leningrad. Studies carried out in Holland and in Hungary in 1944 and 1945 showed clearly that severe malnutrition in large numbers of women results in significant changes in the whole range of vital statistics, from greatly decreased birth rates to greatly increased death rates.

Severe malnutrition is by no means confined to victims of siege and war, and its effects on seriously deprived peacetime mothers and infants have also added to our knowledge. In Philadelphia, a study of infants born to destitute and undernourished women showed a rate of congenital malformation double the rate for the general population. A long-time study in Aberdeen, Scotland, covering seventeen years, showed the same correlation: deficient nutrition in the mother was closely related to malformation in the child.

Another particularly interesting correlation was worked out in the Aberdeen study, between the height of mothers and various difficulties in the birth of their children. The investigators embarked on this study to test their hypothesis that small stature is not only an inherited trait but also, particularly among poverty-ridden people, a result of lifelong malnutrition. They reviewed over 10,000 births and found that the children of women under 5 feet tall were significantly less healthy than those of taller women. Similar observations have been made in the United States on a population of Rhode Island women.

It is ironical that the United States, so much luckier materially than other nations, now has a group of undernourished people whose inadequate diet bears no relation to poverty. This group, notorious beyond the need for statistical proof, is composed of adolescent girls, the mothers of the next generation. Current fashion dictates slenderness, and the girls bow slavishly to this order. Too many adults, failing to understand the full implication of their daughters' malnutrition, allow them to run the risk of real damage to their bodies and to their future children. An undernourished pregnant woman, as we have seen, can counteract much of the effect of earlier malnutrition by beginning to eat properly as soon as she becomes pregnant. But a young woman who has eaten properly all her life and has good dietary habits ingrained in her nature through many years of good nutrition will be, on the whole, better equipped to bear a healthy child. Her own organs will be in excellent condition. They will function smoothly and efficiently and will not break down under the added burden of supplying the new body. And in later years she will pass on to her children the preachments of good nutrition. How much better to start them off, at the most critical period of their lives, with the benefits of it!

The rest of the population could look with indulgence on the folly of these girls, as we do on much of the foolishness of the young, if it were not for the fact that these girls are the very people on whom the destinies of our next generation will depend. Their parents might well ask themselves some searching questions as they watch the girls skip breakfast, eat an apple for lunch, nibble through dinner, and fill up—when they are too hungry to stand it any longer—by wolfing down a chocolate nut sundae at the corner drugstore. Why do adolescent girls eat so much more irrationally than adolescent boys? Boys usually admit to hunger; girls frequently do not. Adults encourage boys to eat, in order to grow strong and manly, to prepare for the role of adult male, or fighter, or competitor. Why are not girls encouraged to eat properly to prepare themselves for their equally certain destiny of childbearing? And why do girls, who when they are small feed their dolls conscientiously and when they are grown feed their families with the same eagerness, neglect to feed themselves? These questions, all related to the problem of the prenatal care of children, deserve attention and earnest study.

The general facts, then, proving the connection between nutrition and the development of children before birth, are well established, gathered from the war-ravaged cities of Europe and the poverty-laden clinics of our own cities. Evidence comes to us, too, from the laboratories, where work with animals has begun to show us specifically what a proper diet is, what the effect of it can be on the development of the young, and what happens when it is lacking. This evidence cannot be transferred, without change, to human beings, but it serves the valuable purpose of pointing out to us the general direction in which we should look in human beings, and in preparing us for what we shall find when we look there.

From work with animals, several investigators have discovered that vitamin A, for example, is essential to the actual process of formation of the organs in early pregnancy. When doses of vitamin A were given to the animals during the period when the major organs were being formed, the young were born normal and healthy. When the mothers were deprived of it, the young were frequently born with underdeveloped major organs and various other defects.

Vitamin B_{12} has also been proved to be essential to proper development. Animals who have been deprived of this vitamin have given birth to young that are suffering from hydrocephaly, or enlargement of the head.

Riboflavin, one of the members of the vitamin B_2 complex, appears to be essential in the development of the jaws, the teeth, and the palate. Drs. Josef Warkany and F. M. Deuschle, who performed the experiments that established this connection, believe that this same vitamin is necessary for the development of these same parts of the human body, and that a deficiency of riboflavin can cause malocclusion of the jaws and teeth and even, perhaps, cleft palate.

Calcium has also been shown to be necessary for the development of the teeth, as well as for the proper growth of the bones. A calcium deficiency has been found to result in rickets, a disease that is characterized by a softening of the bones. Phosphorus and vitamin D are also necessary for good bone construction.

During pregnancy, there is a considerable depletion in the store of iron in the mother's body, because of the need for great quantities of it in the formation of the red blood cells of the child. A diet that is deficient in iron will result in anemia in the mother and in a degree of anemia in the child. Sometimes this does not show at birth, but appears later, after several months. These cases of anemia, in both mother and child, vanish as soon as they are treated with sufficient quantities of iron.

The absorption of vitamins into the bodies of a pregnant woman and her child varies among individuals and even at different periods in a single individual's life. A pregnant woman is frequently more efficient at absorbing vitamins than other women. Likewise, her child often demonstrates the same high efficiency. However, a pregnant woman cannot count on this; many a healthy woman who has been accustomed for years to a normal adequate diet develops vitamin deficiencies during pregnancy. This may be due to metabolic changes in her body, or to the demands of the child, or to a combination of the two. This is a matter that is not yet clearly understood, and a pregnant woman will be wise to be guided by her doctor.

As the pregnancy advances, the mother's efficiency at absorbing vitamins sometimes declines to a point below that of non-pregnant women. On the other hand, as the child grows, he usually becomes more adept at absorbing the vitamins he needs. Similar changes occur in the absorption of iron, in the sugar enzyme content of the blood, and in levels of sugar and insulin. It is as though a ravenous customer with pocketfuls of money had entered the grocery store. The stock is barely adequate for the already established daily needs, there are no supplies in reserve in the storeroom, and the wholesaler is late in his deliveries. The customer has no access to any other store. Clearly, even though he is hungry and greedy, and fully capable of pulling his own supplies down off the shelves, he is in danger of starving.

Thus the nutrition of the pregnant woman must be adapted to meet these changing requirements of her child and herself. The Food and Nutrition Board of the National Research Council has published a table, last revised in 1958, showing recommended daily amounts of the various materials in a good diet. The amounts are designed for the maintenance of good nutrition in people who are already healthy, and the caloric allowances apply to people who are engaged in moderate physical activity; they are too large for women who work in offices or at other sedentary occupations. Here are the figures that apply to pregnant women:

Calories: during first half of pregnancy, 2,300
during second half of pregnancy, 2,600

Calcium: 1.5 grams

Proteins: 78 grams, or about 3 ounces

Iron: 15 milligrams

Vitamin A: 6,000 International Units

Thiamine: 1.3 milligrams

Riboflavin: 2.0 milligrams

Niacin: 20 milligram equivalents

Ascorbic acid: 100 milligrams

Vitamin D: 400 International Units

These amounts are recommended for the average woman, already in good health, with no special problems. Any individual woman who is in doubt about the adequacy of her diet should consult her doctor.

A developing child and his mother both need more than the usual amount of protein, the chief constituent of all the tissues in the human body. During the process of building the child's body, protein is as necessary to him as it will ever be again. Most of the requirements for good nutrition during pregnancy are basically not much different from a good ordinary diet, with the exception of the protein requirement; this is higher in pregnancy than at any other time except during the breast feeding of a child, when it is even higher. If the prospective mother will drink a quart of milk and eat one egg and an ample serving of lean meat, fish, or poultry every day, she will have enough protein under normal circumstances.

Sufficient quantities of calcium and iron will probably be found in a good, well-balanced diet. On the other hand, many women are slightly deficient in iron, although not deficient enough to be labeled anemic; these women will benefit from regular doses of supplementary iron.

Iodine is particularly important during pregnancy, because of the increased activity of the thyroid gland. It is widely recognized that goiter, which is an enlargement of the thyroid gland, is more likely to develop during pregnancy than at other times. Consequently, the use of iodized salt, which is advisable during normal periods, is strongly recommended during pregnancy. Few foods, except seafood, are adequate sources of iodine.

The vitamin requirements, with the possible exception

of vitamin D, are present in a well-balanced diet. Vitamin D, which is necessary in the body to make possible the utilization of calcium and phosphorus, may be supplied by so-called vitamin D milk. Otherwise, it can be taken up as a supplement to the diet.

In addition to the protein requirement in the form of milk, eggs, meat, fish, and poultry, the well-balanced diet includes generous amounts of citrus fruits, tomatoes, and deep yellow and dark green leafy vegetables. Some of all of these should be eaten daily. Butter, cream, margarine, and other fats, starches, and sweets should probably be decreased. The pregnant woman should drink about two quarts of liquids every day, including her daily milk. Five or six small meals every day are preferable to three large ones, particularly during the last six months, when the growing baby will affect his mother's capacity.

The importance of prenatal nutrition has been neatly summarized for us by Dr. Genevieve Stearns, the University of Iowa nutritionist, who has written:

> The best provision for well-being in any period of life is to arrive at that point in good nutritional and physical status. The well-born infant is sturdier throughout infancy than the baby poorly born; the sturdy infant has stores to give impetus to growth in the pre-school years. The child who is in excellent nutrition will have stores to be drawn upon during the rapid growth of puberty. The well-nourished mother can nourish her fetus well; therefore the best insurance for a healthy infant is a mother who is healthy and well nourished throughout her entire life, as well as during the period of pregnancy itself.

IV—THE BREATH OF LIFE

THE MOST critical single element in the life of a child before he is born is his supply of oxygen.

Not food? After we have just finished demonstrating that the difference between a good diet and a poor diet in a prospective mother can mean the difference between good and poor health in her child? Good nutrition is obviously essential to the proper development of every human being. But our need for food, great though it is, is intermittent. We need oxygen steadily, in the right concentration and without interruption, every minute of our lives.

An adult can be killed by the simple act of depriving him of air to his lungs for a very few minutes. Perhaps worse, an even shorter period without air may not kill him, but can damage his brain beyond repair. However, an adult has his own built-in protection against this: when his supply of air is threatened, he fights for it, savagely. Consequently, very few adults die, or are seriously damaged, as a result of insufficient oxygen.

A child before birth, on the other hand, cannot fight for his oxygen; yet he, too, needs it. For this reason he must be carefully protected against accidental deprivation of oxygen, and it is one of the primary jobs of his mother and her doctor to do this.

What his mother can do is the subject of this chapter. In brief, she can learn the basic facts about the dangers that may threaten the oxygen supply of her child, she can review the scientific work that shows the importance of oxygen, and she can put into action the recommendations that have grown out of this work.

Before he is born, a child has no contact with the atmosphere. The oxygen that is so necessary to his survival and to the orderly development of his body reaches him by way of his mother's bloodstream. Her blood, which carries oxygen molecules along with it, bathes the hundreds of thousands of little rootlike villi that extend from the surface of the placenta. The oxygen molecules pass, rather slowly, through the walls of the villi and into the tiny blood vessels within. These lead to the placenta proper, then to the umbilical cord, and finally to the blood vessels in the body of the developing child. The oxygen molecules travel in his bloodstream to his heart, and from there are pumped out through the body to all the growing and developing tissues where the oxygen is needed in order to carry out the work and division of the cells.

When the time of birth draws near, the oxygen level in the placenta and in the child's circulatory system drops sharply, and from that moment until after he is born, he exists and thrives on a relatively small amount of oxygen. At the same time, the placenta stops manufacturing the hormone progesterone, which has until now maintained the pregnancy. Then the uterus begins contractions that will propel the baby down the birth canal and into the air.

The oxygen level of the child's blood just before birth is normally between 50 and 60 percent of saturation. That is, the red blood cells carry on their surfaces only the proportion of the oxygen they are capable of carrying. Two minutes after birth, if all goes well, the concentration will have jumped to 75 percent. Four minutes later it will be 90 percent saturated, and when he is thirty minutes old, his blood will be 95 percent saturated with oxygen. And from that moment on, for the rest of his life, he will need this much oxygen for his best health.

Although the fetus does not breathe, there is evidence that its chest moves rhythmically in a typical respiratory pattern. Even so, the establishment of independent breathing in a newborn child is one of the many marvels of the human body. When mothers and doctors allow it to proceed naturally, the complex actions and reactions of the various forces at work flow smoothly along from beginning to end. For example: every contraction of the uterus momentarily interrupts the flow of blood to the child. This, of course, deprives him for that brief time of all oxygen, and oxygen starvation—or *anoxia*—might be the next step. Normally, however, anoxia does not develop, because at the same time that this is going on, the child's temperature is dropping. This lowers his rate of metabolism, which means that for the moment he actually needs much less oxygen in his blood. Thus, what might be a threatened anoxia is avoided by a compensating mechanism.

Sometimes, unfortunately, this natural process is interfered with, and the results are not always good. Recent studies have shown us that several practices that were common and believed at one time to be beneficial can be dangerous to the child who is being born. A mother who permits them is risking serious damage to her baby. The use of an improperly given general anesthetic is one of these practices. During the actual process of birth, the doctor, of course, is in charge, and no mother in labor has time or inclination to discuss these matters with him then. But when she chooses her doctor in the first place, early in her pregnancy, she would be wise to find out what his views are on the use of anesthetics and analgesics and other drugs during labor, and to choose a doctor who is aware of the work that has been done in this field and the importance of it to her child.

Doctors sometimes give a patient a drug to promote rapid labor. In some cases, of course, this is necessary, but in all cases it carries with it the possibility of a threat to the child's supply of oxygen. One effect of these drugs is to lengthen the duration of each contraction of the uterus. This accomplishes the result of bringing the baby to birth sooner, but it also lengthens the periods when the child is receiving no blood, and hence no oxygen, from his mother. An eminent obstetrician, Dr. Nicholson J. Eastman, of the Johns Hopkins Hospital, has expressed himself forcefully against this use of drugs: ". . . if the contraction should last five minutes (as sometimes happens, alas, from the injudicious use of oxytocic drugs) it is well known that the fetus dies forthwith from anoxia." The important words here are "injudicious use." These drugs sometimes must be used, for the protection of both mother and child, but an informed and careful doctor will use them with the greatest caution. And a wise mother will know well ahead of time whether or not her doctor is informed and careful. A further discussion of the use of drugs and their effect on the child before birth will be found in Chapter VI.

A prospective mother should also know her doctor's views on anesthetics and analgesics during labor. A general anesthetic is probably the most common danger to children during birth, particularly in the United States, where it is still widely used. There is very little doubt now, in view of investigations and studies in this field, that all anesthetics and most sedatives and pain-relieving drugs reduce the oxygen content of the blood. In addition, barbiturates, morphine, and similar drugs have a depressing effect on the mother's own breathing. This reduces the oxygen in her blood, and the supply to her child is cut down in still another way.

In cases where the child is premature, this can be very dangerous, and it has been known to result in the death of the child. A premature child is not so ready in his own body to establish independent breathing and circulation, and when his difficulties are increased by a shortage of oxygen, they are sometimes too much for him. Dr. Arthur G. King, of Cincinnati, investigated a group of premature children who had died during the period immediately before birth or during or immediately following birth, and he came to the conclusion that nearly 30 percent of the deaths were a result of anoxia caused by the anesthetics and analgesics given to the mothers during labor. He said:

> We felt that analgesia-anesthesia had interfered somehow with the mechanism the newborn infant needed to develop its own means for acquiring oxygen in its blood-stream. In going over the records, we were amazed at what we had been using: tremendous quantities of narcotics and barbiturates and some very prolonged anesthetics.

A specific recommendation for anesthetics with premature births has been made by two anesthesiologists of Hartford Hospital in Connecticut. Dr. David M. Little, Jr., and Dr. William K. Bannister suggest that:

> . . . medication be restricted to the minimum compatible with the maternal pain threshold, and that some form of regional anesthesia be chosen: . . . These should spare the premature infant from further depression and are much preferable to the general anesthetic agents.

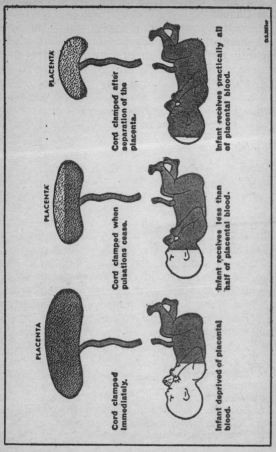

The placental blood the baby needs and clamping of the cord. Showing the amount of placental blood passing to the baby when (1) the cord is clamped immediately after breathing is initiated, resulting in the baby receiving none of the placental blood; (2) when the cord is clamped after its pulsations have ceased, the baby receives less than half the placental blood; (3) when the cord is clamped after the separation of the placenta, the baby receives practically all of the placental blood.

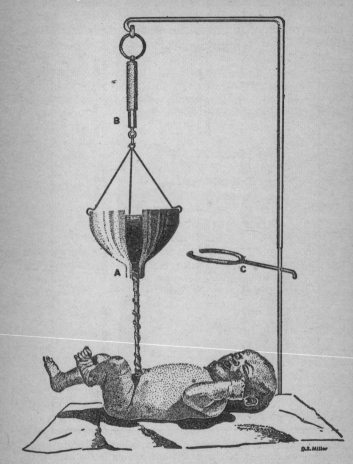

Technique of draining placental blood into cesarean-delivered babies, which may also be used in other cases. (A) Placenta in funnel with blood flowing through umbilical vein into baby. (B) Spring-balance from which funnel hangs. (C) Ring on which funnel rests while placenta is put into it, after which the ring is turned aside. The transfused placental blood volume is read from the spring-balance.

Dr. Eastman, of Johns Hopkins, has also made a recommendation: that oxygen be routinely administered to mothers during the last five to fifteen minutes before delivery in order to protect the welfare of the occasional baby who may be suffering from some degree of anoxia.

Another recent development in delivery-room procedures is the recognition of the importance of the umbilical cord in supplying the newborn child with blood. In the past, the cord was commonly clamped or tied immediately after the birth of the child. But the cord contains a large amount of oxygenated blood, and the child needs this in order to function at his best during the first moments of his life. He should be allowed to have it. Dr. Q. B. DeMarsh and his colleagues, who have studied this matter quantitatively, say that this blood can prevent anoxia in newborn infants. These investigators found that when a cord was clamped immediately after the birth of the child, he was deprived of all the blood in the placenta. When they waited to apply the clamp until after the pulsations in the cord had stopped, the baby still received less than half of the blood from the placenta. When they waited from five to twenty minutes, until the placenta separated, the baby received almost all the blood.

The practice in some hospitals now is to place the placenta, or the "afterbirth," above the level of the baby, so that the blood will flow through the cord into his body. In several hospitals in Europe, devices are in use that milk the blood through the cord. In about an hour, the cord shrinks and dries up. At that time, it is easily cut, and does not even need to be tied, since the blood vessels will have shrunk considerably by this time, and be almost closed. This technique offers an additional advantage, too: it greatly reduces the danger of infection in the newborn child, which occasionally followed the earlier practice of clamping or tying the cord.

But doctors cannot be asked to assume all the responsibility for supplying a child with sufficient oxygen. Recent studies have focused the attention of prospective mothers on their own habits and attitudes, not only during the period of birth but all during pregnancy. It is now established beyond question that smoking reduces the amount of oxygen in the blood of the smoker, and that when a pregnant woman smokes, she is cutting down the amount of oxygen that is available to her child. A study of over 2,000 English women, for example, has shown a significant connection between the number of cigarettes smoked by a mother and the weight of her baby at birth. The babies of heavy smokers weighed least of all the children studied, those of moderate smokers came next in the scale, and the children of nonsmokers weighed the most. Dr. C. R. Lowe, who conducted the investigation, suggested that since it is known that tobacco smoke will constrict the blood vessels, it is reasonable to take this a step further and conjecture that the child's blood vessels may also be constricted every time his mother lights a cigarette. Twenty or more such constrictions every day, all during pregnancy, would clearly reduce the amount of oxygen that would reach the child's tissues, and this, in turn, would affect his growth. The old bugaboo with which parents used to try to frighten children away from smoking, that it would "stunt their growth," turns out, after all, to be quite true, especially when it is applied to the smallest children of all. It does indeed stunt their growth.

Smoking has also been correlated with the premature birth of children. In numerous studies, investigators have shown clearly that large groups of women who smoke heavily have many more premature babies than groups of women who smoke only mildly or not at all. The reasons for this have not been established yet: tobacco smoke is a complex substance, and it contains several chemicals that can affect cell division and other changes that occur in a growing child; these are being studied now, but no definite conclusions have been drawn. In any case, there is no doubt that a premature child runs more risks during his first months of life than does a child who is born at term, and anything that contributes to prematurity, even if it is not entirely understood, should be avoided if possible. A fuller discussion of the effects of smoking on children before birth will be found in Chapter VII.

Drugs bought casually over the drugstore counter and taken by a pregnant woman without consulting her doctor can also play an important role in her child's life. There are several reasons for this, which are also taken up at length in Chapter VI, but one of the most vital has to do with the oxygen content of the blood of both the mother and the child. Many drugs act in the bloodstream just as anesthetics do, to reduce the amount of oxygen and to depress the mother's whole respiratory system. By far the most sensible recommendation on this matter is also the simplest: a prospective mother should not take any drugs at all during pregnancy unless they are essential, and even then only under the direction of a doctor.

In general, the best possible insurance against oxygen starvation in a child before birth is a thoroughly healthy mother who lives a healthful life during her pregnancy and declines to tamper with the natural processes. This means, of course, that she must eat properly. It means that she must rest enough so that she is not exhausted. The amount of rest she needs will depend on the details of her life, and they will vary widely with different women and with different times in each woman's pregnancy. She must give this matter careful attention, for if she becomes too tired, her whole body will suffer, her circulation will be affected, and her child will feel the effects of it in the amount of oxygen that comes through to him. This is no time for a woman to try to be a Spartan. Illness, too, like fatigue, is a drain on the body's resources, and she must avoid illness as much as possible.

If she does all these things, her body will perform its natural functions with very little conscious help from her. Her bloodstream will carry along with it the nutrition and the oxygen that she needs both for her own body and to pass on to her child so that he may build his body. The placenta, which needs oxygen as much as any other organ while it is in the process of growing, will develop in its orderly way and thereafter take over the vital role it must play in transferring nourishment and oxygen from mother to child. With enough oxygen in his own bloodstream, the child will then develop, tissue by tissue and organ by organ, on the schedule that is built into his cells. And finally, when the time comes, the oxygen will withdraw from the placenta, the contractions will start, and the child will be born, healthy and thriving.

To see most clearly the importance of the mother's general health in producing a healthy baby, we have only to look at groups of women whose general health is poor, and then examine their children. The view is disheartening: poor health in the United States is most often found where poverty and ignorance combine, in our city slums. When we investigate the newborn white children in these areas, we find anoxia as a factor in their birth history three times as often among these children as among children born of parents with more money and more education. And we find an anoxic history *ten times* as often among poor Negro children as among middle-class white children.

These figures are revealing. They demonstrate that anoxia is most often a result of the complex set of conditions that afflict women whose whole lives have been a long series of deprivations. Such women have never in their lives had an adequate diet, even during their own prenatal days, and as a result their entire bodies function poorly. They eat no better during pregnancy than at other times, and their children grow with less vigor than more fortunate children. Furthermore, they may develop a flaw here or a defect there because of an insufficient supply of oxygen at a critical moment. And it is certain that the whole double circulatory system—mother's and child's, connected by the placenta—is less efficient in such a pair than in a thoroughly healthy mother and child. The amount of blood may be less, or the quality of it may be poorer, or even both. Oxygen is transmitted in smaller quantities and is distributed throughout the child's body more sparingly. This is true oxygen starvation, and its consequences can be serious.

The most serious effect of anoxia to a living child is brain damage, and the studies that have been carried out on the relationship between anoxia and various kinds of damage to the brain are among the most hopeful that have come out of the laboratories in recent years.

The hope stems from this: brain damage that is a result of a lack of oxygen—and much of it is—can be prevented. The importance of this conclusion can hardly be overemphasized.

Any kind of brain damage is serious in itself, because it is both quick and permanent: the cells of the brain are the most vulnerable of all the cells of the body to a lack of oxygen; they can be destroyed more quickly than any other cells. Also, when they are destroyed, they are gone forever, because they cannot regenerate themselves as cells of the skin do, for example, after a cut or a laceration.

There is no healing process of this kind in the brain. Furthermore, the brain controls the human being, and when a portion of the brain is damaged, the part of the individual that is controlled by that portion of the brain is also damaged. A number of mankind's more serious maladies are sometimes brought about in just such a way.

Cerebral palsy, for example, is a disorder that develops when the part of the brain that governs the muscles and muscular coordination is damaged. Epilepsy is the result of damage to the nervous system. Mental retardation is a sign that something has interfered with the proper development of the individual's thinking apparatus. And some mental illness is now thought to be a result of the faulty development of a part of the brain. All of these conditions are sometimes caused, according to investigators, by a lack of oxygen to the brain.

Not all cases of these conditions are a result of anoxia, obviously. And if it were possible by some miracle to arrange sufficient oxygen for every child before birth, there would still be cases of cerebral palsy among us, and epilepsy, and mental retardation from other causes. But the numbers of them would be greatly reduced, and the human race would be immeasurably the richer for the presence of these people, undamaged.

There is some evidence that even mild anoxia may cause some slight damage to the brain. Some reading difficulties, for example, are now believed to have their source in brief periods of oxygen starvation before or during birth. And some behavior problems may be traceable to the same origin.

But how do we know these things? What kind of studies result in such conclusions as these?

A good example of such a study was carried out by two investigators who have spent many years of their lives in working out techniques for discovering valuable information in the medical histories of human beings: Dr. A. M. Lilienfeld and Dr. Benjamin Pasamanick. Their subjects were 600 children in New York City who had cerebral palsy. The goal of the investigation was to determine, if possible, what events in their lives or in the lives of their mothers before the children were born might have caused the cerebral palsy, with the ultimate purpose of learning how to prevent such damage in the future.

The investigators went about their study by looking specifically for conditions that produce a lowered oxygen content of the blood. How many of the children had been premature? they asked. And how did this proportion of the whole group compare with the proportion of children in the general population who were premature? It is well known that premature babies frequently have trouble breathing—their bodies are not quite ready to function independently—with the result that many of them get too little oxygen. The doctors found that four times as many of the palsied children had been premature as had children in the general population. This means that prematurity and cerebral palsy are undoubtedly connected, and the link between them may be a lack of oxygen. A premature baby: not enough oxygen: damage to the brain: cerebral palsy. This is one possibility.

Another connection was found between complications of pregnancy and cerebral palsy. Complications of pregnancy are a group of specific maladies that frequently result in a decrease of oxygen in the blood. They were found to have occurred in 38 percent of the mothers of the palsied children. This is far more often than the incidence of such complications among women in general. The linkage in these cases may have gone thus: complications of pregnancy: not enough oxygen: damage to the brain: cerebral palsy. This is another possibility.

In some cases it may well be that the lack of oxygen produces not only cerebral palsy but also prematurity, while in others the prematurity is responsible for the cerebral palsy, and in still other cases the cerebral palsy results in prematurity.

A misplacement of the placenta, called placenta previa, which owing to bleeding affects the amount of oxygen that can be transmitted from mother to child, was also found to be frequent among these mothers. Placenta previa: lack of oxygen: brain damage: cerebral palsy.

However, if we are trying to prevent cerebral palsy in human beings—and our medical research workers are certainly at least trying to find out how it might be possible to prevent it—we must go back a step farther still. We must ask what caused the prematurity, and the complications of pregnancy, and the placenta previa. They, apparently, caused the anoxia, which in turn caused the palsy. But what caused them?

And here we are back again where we started: the good health of the mother, arising generally from a life of healthful living and primarily from a good diet. Women in good health have far fewer premature babies than women in poor health. They have complications of pregnancy much less often and less severely. And they have strongly functioning organs with which to nourish their children with both food and oxygen.

None of us, as individuals, can hope to wipe out completely the maladies of mankind. But thanks to the work that is being done by such men as Dr. Lilienfeld and Dr. Pasamanick, the prospective mother can know a great deal more about the causes of these illnesses than her mother and grandmother knew, and she can take the simple precautions that arise from this knowledge.

Epilepsy is another condition that is sometimes brought on by a lack of oxygen, and that a prospective mother can therefore help to prevent. The same doctors who did the work with cerebral palsy traced a similar connection between anoxia and epilepsy. They studied over 400 children this time, all of them epileptic, and compared them with the same number of healthy children. Again they looked into the proportions of both groups that had been premature, and whose mothers had had complications of pregnancy. Again they found that the figures were much higher for the epileptic children than the healthy ones. And this time they looked specifically at the birth records of all these children for mention of cyanosis. This is a blue or purplish color of the skin that indicates a lack of oxygen in the blood. Cyanosis was found on the records of very few of the normal children, but on the records of many of the epileptics.

Among these children, then, the sequence of events probably went like this: prematurity (or complications of pregnancy): lack of oxygen: brain damage: epilepsy. This time, instead of affecting the part of the brain that controls muscular coordination, the anoxia permanently altered the structure of the nerve cells of the children; hence epilepsy instead of cerebral palsy.

A lowered intellectual capacity is also sometimes connected with anoxia at birth. This is a particularly difficult connection to trace. How can an investigator possibly say that any given child would have been brighter if he had had more oxygen in his bloodstream? He can't. But he can measure the intelligence of large groups of children who are known to have had anoxia at birth, and other large groups who are known not to have had it, and compare the scores of the two groups. Such a group in St. Louis is being studied in a long-term investigation, and although the work is not yet finished, the conclusions seem clear enough. The two groups of children overlap, of course; many anoxic children are brighter than many children who did not have anoxia. But the middle I.Q. score of the anoxia group—what is called the *mean*—was seven points lower than the middle score of the other group. This same difference was found in vocabulary tests. Furthermore, the children who had been only mildly anoxic at birth scored higher than those who had been severely anoxic. Tests were also made to assess the neurological health of these children, that is, the healthy functioning of their nervous system. Most of the children of normal birth were characterized by a normally functioning nervous system, whereas most of the anoxic children were not.

A prospective mother can apparently even help her child to learn to read better, by ensuring the supply of oxygen to him before he is born. Dr. Benjamin Pasamanick and Dr. A. A. Kawi made the connection between anoxia and reading difficulties when they studied over 200 boys in the reading clinic of the Baltimore Department of Education and compared them with the same number of boys who were normal readers. The birth records of the normal readers showed that only 1.5 percent of them had suffered from any of the complications of pregnancy that result in reduced oxygen. On the other hand, nearly 17 percent of the poor readers had been anoxic. On reflection, this seems reasonable: in adults, an injury to the brain sometimes interferes with the ability to read; the same kind of injury to a child might well result in the same kind of effect. A child who has difficulty with reading, therefore, is not necessarily "stupid," but may have suffered from anoxia long before he ever saw a book.

A mother can even have an effect on her child's subsequent behavior through her attention to his oxygen supply. Many children who are behavior problems might not be so if they had had sufficient oxygen before or during their birth. The lack of it may have caused a certain amount of brain damage that is expressed in uncontrolled behavior. More than a thousand children in Baltimore who suffered from behavior disorders were studied by Dr. Pasamanick, to try to determine whether brain damage was more common among them than among children whose behavior was acceptable. His conclusion was that the mothers of these children had had far more than the normal proportion of difficulties during pregnancy, and that many of these difficulties were accompanied by a lowering of the oxygen content of the bloodstream. Thus it seems quite likely that many children whose behavior is unacceptable are not simply "ornery" but are the victims of actual damage to their brains.

Dr. Maurice W. Laufer and Dr. Eric Denhorf have called one group of behavior symptoms "hyperkinetic impulse disorder," and they have found that many of these children have damaged brains, and that some of the damage was done by anoxia. Children who suffer from this disorder are hyperactive, have a short attention span and poor powers of concentration, and are impatient, irritable, and explosive. With these traits, they fit very poorly into an average school classroom, although their intelligence may be perfectly normal. The symptoms disappear spontaneously as the child grows up; the investigators saw no person with these symptoms who was over eighteen. By this time, however, the child has encountered so much difficulty in his life, from his teachers and his parents, that he may have developed another set of defense symptoms that can disable him for years. For this reason, even though this disorder is not permanent, it must be considered a serious problem for many children and adolescents.

Mental illness, too, is controllable to a certain extent

in this same manner. Several mental disorders have been shown to appear much more frequently in individuals who have had anoxia than in those not born anoxic. The mechanism of this may be that the central nervous system is damaged by the anoxia while it is developing, with the consequence that in later years it is more susceptible to the stresses of life. The nervous system of the healthy child is much stronger and resistant to strain.

The same mechanism might apply to the development of the heart and blood vessels of a child. If these develop healthily, with all the nutrition and oxygen they need, they will be able to supply the brain and nervous system of the individual with enough oxygen all during his life. But if a child's heart and blood vessels develop poorly, because they lacked oxygen at a crucial period during his prenatal life, they will never function entirely adequately, and his brain will never get quite enough oxygen. This in turn can contribute to mental breakdown.

A prospective mother who wants the best possible life for her child will remember all these connections. She can influence her child's development toward health most readily by conducting her own life during pregnancy so that she is not exposing him to these threats.

She can do this by following these recommendations:

1. She should check with her doctor early in her pregnancy to be sure that she knows and agrees with his views on anesthetics, analgesics, and other drugs during labor.
2. She should not have more than a regional anesthetic during labor.
3. She should not take any drugs during her pregnancy unless they are essential, and unless she has her doctor's approval.
4. She should find out what her doctor's practice is with regard to the clamping or tying of the umbilical cord.
5. She should not smoke.
6. In general, she should live as healthfully as possible, with a good diet and enough rest.

If she does these things, which represent the best and

most recent recommendations based on the findings of our medical investigators, she will be doing all that one private individual can do to reduce the number of children in the next generation who suffer brain damage from anoxia. In addition, she will be safeguarding her own child.

V—A MOTHER'S AGE

Is it better for a child to have a young mother than a middle-aged mother? Is a girl of eighteen physically better qualified to nurture her child than a woman of thirty-eight? In short, is the growth and development of a child before birth at all affected by the age of his mother? And if the child is affected, how does he show it?

These are some of the questions that have been asked and at least partially answered during the course of recent investigations into the health and development of children before birth.

These investigations are not so difficult to carry out and interpret as some others. Large groups of women and their newborn children are observed, either directly or through their hospital records. The weight and the condition of each child and the age of his mother are recorded. When the child is normal and healthy, which he is in the overwhelming majority of cases no matter what the age of his mother, this fact is quickly noted. If the child is not healthy, if he was born prematurely, if he was anoxic at birth, or if he has not developed properly, these facts and the relevant details, plus the age of his mother, are also noted. Her experience is also part of the record: whether she had any difficulties during pregnancy, how long labor continued, what kind of delivery it was. When the study is finished, certain general statements can be made about this particular group of women and children. If it is a very large group, or if the conclusions agree with other studies that have gone before, it begins to be possible to make general statements about the population as a whole.

For example:

1. Mothers between the ages of twenty-three and twenty-nine have the largest proportion of healthy, normal children.

2. Younger mothers, between seventeen and twenty-three, run a slightly greater risk of having premature or stillborn children, or children who have not developed properly.

3. Some mothers between thirty and thirty-five, particularly those who are in poor health, begin to show the results of aging in the condition of their children at birth.

4. After thirty-five, the risks to the child increase more sharply.

The problems posed by the studies on the age of mothers are found chiefly not in setting them up, or in reading the results, both of which are simple enough for the experts, but in applying their lessons to individual women. Statistics make no predictions about individuals. Plenty of girls of eighteen and women of forty-two have had children who were magnificently normal and healthy. And sometimes, owing to causes other than age, a sickly, undeveloped child is born to a woman who is in the middle of her most propitious age span.

One of the difficulties for an individual woman is in understanding that chronological age and biological age are not always the same. Her chronological age is simple to find: how many years have elapsed since she was born? Her biological age is more difficult to discover. This is determined by the rate at which her body ages.

Aging is a continuous process: it begins at conception and goes on every moment until death. In the early part of the life cycle, the cells of the human body grow and divide and develop. The organs and the structures of the body increase in complexity and strength and efficiency. Finally they arrive at the peak of their development. Then, after a time, the body slowly begins to lose its efficiency and its strength.

The difficulty lies in the variation of the rate of this aging process from one person to another. We all know elderly people, for example, who appear to be much younger than other people who were born in the same year. Their muscles are more flexible, their skin more taut, their ailments less serious, and their minds more active. This variation holds true with respect to women and their ability to nurture their children. Some women are fully mature in all the functions of their bodies by the age of twenty; the majority reach it at about twenty-three; still others are not mature until twenty-seven.

The aging process varies, too, within each person, from organ to organ, from system to system. We have become accustomed to the medical description of the man with a heart older than the rest of his body. In just such a fashion there is variation within each woman with respect to her reproductive system in contrast to the rest of her body.

And this is not all. When a prospective mother inspects these studies on maternal age, she naturally wonders in what other ways she differs from the women who were studied. The rate of aging may well be affected, at least partially, by an individual's general health. No studies on this subject are available, but we can be convinced of it when we think of the gaunt, undernourished women in our southern mountains, who give the appearance of being fifty years old when they are still under thirty. Such a woman is less likely to bear a healthy child, at whatever age she is, than a woman much older whose general health has always been good.

And how many children does she already have? There is a widespread belief among women that too many pregnancies will exhaust a mother physiologically and that a child who comes at the end of a long line is apt to suffer from his mother's exhaustion.

This is a difficult subject to study, because it is not always possible to separate the factor of the mother's age, from the factor of how many children she has already borne. Obviously, the more children she has, the older she is, but if her next child is less healthy than the preceding ones, which is the reason, her age or the number of her previous pregnancies? Or both? Numerous studies have been attempted on this subject. These will be discussed later, and some of them appear to lead to logical conclusions.

Fatigue must be considered, too. A woman who has had five children within eight years, who is taking full care of them, and keeping house, and cooking and doing the laundry for the crowd, is most certainly a tired woman. She works very hard for long hours at a job that is emotionally demanding as well as physically exhausting. This kind of fatigue may affect the development of her next child. It is a mistake, however, to say that either her age or the number of her previous pregnancies is the cause. Even with a houseful of small children, she will not be too tired if she has sufficient help so that she can get enough rest. If she is basically healthy, eats sensibly, if she likes her job and is reasonably relaxed about it, the chances are overwhelming that her next child will be perfectly normal and healthy, even if she is thirty-eight and the mother of five.

In other words, although statistically speaking, there is a period when women taken as a whole are best fitted to bear children—between the ages of twenty-three and twenty-nine—these figures have only a limited significance for any individual woman. They cannot possibly predict the health and development of her coming child. Their chief value for her is to teach her something about the workings of her own body, to show her that a period does exist when her body is probably at its best, and to guide her in the care she takes of herself and her child at whatever age she is.

In this era of early marriages and early motherhood, the most startling and important result of the studies on maternal age is perhaps the discovery that too young can be as risky as too old.

The development of a woman's entire reproductive system occurs step by step, and it involves a great deal more than the simple ability to conceive. The ability to nourish and protect a child for nine months is a more complex requirement; it takes longer to achieve it. It is not surprising that some very young pregnant women are not quite ready.

There are records, of course, of extremely young mothers who not only conceived children at a much earlier age than usual, but carried the children successfully until birth. The most startling of these was Lina Medina, of Peru, who was only a small girl herself, five years and eight months old, when she gave birth in 1930 to a healthy six-pound boy. He has since grown up, normal and intelligent, and when last heard of was planning a career in accounting. There are numerous other similar, if not quite so extreme, cases described in the pages of medical journals.

Nevertheless, the statistical fact remains that mothers over twenty-three have a better record, for themselves and their children, than mothers under twenty-three. They have fewer miscarriages, premature children, stillborn children, and children who are ill or in some way undeveloped.

The reason for this is not hard to find. There are two well-defined stages in the development of a girl's reproductive system before she reaches maturity: puberty and nubility. These words do not mean the same thing, although they are frequently used as though they did. Puberty is the period after a girl has begun to menstruate but before her ovaries have begun to function. Nubility commences when the eggs begin to pass out of the ovaries. After that comes the period of reproductive maturity.

During puberty, a girl is not usually able to become pregnant; she is sterile. She is developing the secondary sex characteristics—the broadening pelvis, the breasts, the pubic hair—but even though monthly bleeding occurs, her ovaries are not yet producing the egg cell, or ovum, that is necessary for the start of a new life. The endocrine system of glands in her body must achieve a distinct level of development before this can happen. When the glands are sufficiently developed, the ovaries can produce the eggs.

This step in her development is nubility. During this period, although she can become pregnant, she is not yet mature. Nubility, in fact, represents a bridge between puberty and maturity. During nubility, several of the endocrine glands undergo various definite changes that gradually, step by step, act on each other and on the ovaries until an entire series of steps has been concluded. Near the end of this series, the pituitary gland releases a hormone known as luteotrophin. This hormone, reaching a particular spot on the ovary, acts on it and causes it in turn to release another hormone called progesterone. This word comes from two words meaning "in favor of" and "gestation." The function of progesterone is to prepare the uterus for pregnancy. When this step has been taken, a woman's reproductive system is mature.

Thus, if she becomes pregnant before the endocrine glands have completed their work, the fertilized egg will be deposited in a uterus that is not yet quite ready to receive it. The egg can grow and develop there, and in by far the majority of cases, the child will be a normal, healthy child, but the odds in favor of this will not be so great as they will be later, when the reproductive system is altogether ready.

A mother who has not yet completed the development of her reproductive system, then, runs more of a risk with the health and development of her child than the mother who is mature. Furthermore, she runs another risk: that of being herself psychologically too immature to take on the job of motherhood. There is enormous variability in this quality among girls and women, but there is no question that very young mothers are more frequently inadequate mothers than older women. And until it has been shown to be false, a reasonable supposition is that development of the entire body and its physical systems marches along in any human being with the development of emotional and psychological maturity.

In the long run, then, early motherhood is probably a disadvantage to both mother and child, and should not be encouraged. All the evidence strongly suggests that twenty-three is a much better time, statistically speaking, to have a child than eighteen.

The peak of maturity in a woman's reproductive system lasts about five years. This appears to be the case whether she is early, average, or late in her development.

After the period of maturity, a gradual process of aging begins to set in. Her reproductive organs slowly grow less adept at their jobs. This is expressed, for example, in the birth weights of children of women of various ages. The children of women from twenty to thirty, according to a study of over a thousand consecutive births in India, varied in weight as the age of the mothers increased: the lightest were born to the youngest women, and the heaviest to the oldest. After the age of thirty, the birth weights decreased gradually but consistently as the age of the mother advanced. The weight of a child at birth is a clear indication of how well he has grown during his prenatal period; generally speaking, the heavier the child, the healthier he is. Therefore, a decrease in birth weight over a large number of children as their mothers grew older means a decrease in their nourishment and general health. Other evidence appears to corroborate this: at the end of the first year of life, infants of young mothers are taller and heavier than those of older women.

The mechanism of the aging process in general is not well understood, but some aspects begin to be clear from the studies of large numbers of women and newborn children. The large majority of women, of course, continue to bear healthy children as long as they continue to have children at all. Some few women, however, whose bodies have never been particularly healthy, begin to show the effects of their poor nutrition and other difficulties in the aging of their reproductive systems. These mothers begin to have more babies who are stillborn, or premature, or poorly developed than they had in their earlier years. And they begin to show their aging in their own physical problems during pregnancy, in more cases of prolonged labor and more toxemias of pregnancy.

At first only the seriously deficient mothers are affected in this way. A woman of thirty-three, for example, who has always been healthy and continues her good habits of eating and exercising and resting, cannot seriously be said to be growing old. But a woman of thirty-three who is sickly, frail, undernourished, and in poor health generally, *is* growing old, in her reproductive organs as well as in her other systems and in her appearance.

The decade between thirty-five and forty-five finds more women whose bodies are not quite equal to the task of adequate reproduction. Among these women, some whose health has been moderately good are caught between their deficiencies and the aging that happens to everyone. Some of these women give birth to premature children, or anoxic children, or stillborn or defective children.

Part of the reason for this lies in the quality and the quantity of the egg cells in the ovary. At birth, a human

ovary contains some 200,000 egg cells; these steadily decline in number with the passing of the years until, at
about the age of fifty, there is not a single egg cell left in
the ovary. Some of these have developed into mature eggs;
others have broken down and disappeared. Furthermore,
the gradually diminishing number are also growing older.
They are less active than they once were.

In the meantime, the ovary itself is becoming thinner,
smaller, and more wrinkled. The female sex hormones,
the estrogens, themselves help in this aging process, because they tend to stimulate the growth of fibrous connective tissue in the ovaries and uterus. In fact, by the
age of thirty, nearly a third of all women have developed
fibroid tumors in the uterus; by fifty, these tumors have
appeared in half of the female population. The uterine
tubes, the vagina, and the breasts all undergo similar
changes—loss of function, slowing down of activity, and
lessening of elasticity. All these changes, occurring slowly
over the years, gradually produce a less hospitable environment for a developing child. And it is these changes
that may be responsible for the occasional stillborn or
premature or defective child and for the statistics that add
up to the greater risk that exists in the lives of children
of older women.

One of the most interesting and startling developments
in the study of the age of mothers was the correlation,
made in 1959, between maternal age and Down's syndrome, formerly called "Mongolism." This condition,
which is a form of mental retardation, was first described
by J. Langdon Down, an English physician, in 1866.
Down termed these children "Mongolian idiots" because
to the untrained Caucasoid eye they seemed to resemble
the Mongoloid peoples.

After nearly sixty years of speculation about the cause
of this disorder, the subject was still so little understood
that it was possible to publish a book in 1924 entitled
The Mongol in Our Midst, in which the theory was seriously advanced that the occasional appearance of the
"Mongolian idiot" was a "throwback" to the Mongoloid
ancestry of European peoples. Actually, the resemblance
is due simply to the development of a fold in the upper
eyelid which is not usually found among Caucasoids, but

which is caused not by inheritance but by the failure of the root of the nose to develop normally. Ironically, the Japanese profess to see in their children born with this disorder a resemblance to Europeans. It is desirable for scientific as well as for social reasons that the term "Mongolism" be dropped in favor of "Down's syndrome." This would follow the tradition of naming a disease or disorder after the person who first described it.

Children with Down's syndrome are born in all segments of the population, to the poor, the rich, the well-nourished, and the undernourished, and to those of all hereditary backgrounds. They appear in all ethnic groups —American Negro, Asiatic Indians, Chinese, and Japanese, as well as Caucasians. All these factors were studied in the long and slow investigation into its causes. The answer to the puzzle was not found until the ages of the mothers were correlated with the birth of children with Down's syndrome.

In 1932, Dr. C. Bennholdt-Thomsen published figures that showed this relationship: As the ages of the mothers increased, the frequency of Down's children in the group rose. After thirty-five, the rise was considerable; after thirty-nine, it was precipitous. Among a thousand children of mothers between twenty and twenty-four, there was only one Down's child. Among a thousand children of women between forty and forty-four, there were more than forty Down's children. Among a thousand children of women between forty-five and forty-seven, there were more than 120. Upward of 7,000 Down's children are born annually in the United States, and the great majority of these are born to mothers over the age of thirty-five.

A partial explanation of this was offered in 1934 by Dr. Adrien Bleyer, who suggested that as women grow older the smooth, automatic functioning of their ovaries becomes impaired. He speculated that the chemistry of the egg cells might be disturbed with age, which might result in changes in the chromosomes of the germ cells. Chromosomes are the tiny, threadlike structures that lie in the nucleus of a cell and carry the human hereditary material in the form of genes. Twenty-five years later, in 1959, Dr. Bleyer's theory was confirmed by Dr. Jêrome Lejeune and his colleagues of Paris, in a careful investigation of nine cases of Down's syndrome, each of which proved to have forty-seven chromosomes instead of the normal forty-six. Dr. Lejeune believed that this was the result of a faulty development of the egg cell.

Another explanation has been put forward to explain the correlation of maternal age and the incidence of Down's syndrome: this concerns the aging of the ductless glands in the mother's body. The hormones that are secreted by these glands are necessary for the proper development of the egg cell, and when they fail to appear, or when they appear in insufficient quantities, the effect on the child may well be to retard his growth.

There is considerable hope among medical investigators that future research into hormones will show a way in which prospective mothers can be tested for the adequacy of their hormone secretion. When this is accomplished, it is quite possible that a woman whose glands are not sufficiently active to supply her developing child may be given the hormones she needs. Thus the number of children with Down's syndrome born to mothers of all ages may be dramatically reduced.

Many authorities now believe that Down's syndrome may be a result of a combination of these two factors: an aging of the ovaries themselves, which results in chemical changes in the egg cells, and an inadequacy in the secretion of various hormones. It is quite possible that both of these situations need to be present in order to produce Down's syndrome.

Maternal age also seems to affect the frequency of births of twins. Physiologically speaking, the birth of twins can be called an abnormal event; it is due either to the splitting of a single egg, which produces identical twins, or to the fertilization of two separate eggs, which results in nonidentical or fraternal twins. The age of the mother apparently affects only two-egg, or fraternal, twins; one-egg twins are born with equal frequency to every age group. When a woman's ovary sheds two eggs instead of the usual one, the reason may lie in the fact that it is growing older and less efficient.

Two-egg twins occur least often in young mothers, and then their incidence increases progressively with a mother's age until she reaches thirty-eight. After this, the proportion drops sharply. The reason for this is not clear, but the suggestion has been made that when an older mother sheds two eggs, one of the twins dies in the uterus, at an early stage of development.

Twinning also increases with parity regardless of the mother's age. The more children a mother has, the greater the chances of twinning, up to and including the ninth pregnancy. Thereafter the chances of twinning decline.

The ratio of boys to girls is another area in which the age of the mother appears to have some effect. Among the white population of the United States, in general, there are 106 boys born for every 100 girls born, and as far as we know, this ratio is the same in other parts of the world. However, the ratio is not the same for all age groups. In the younger groups of mothers, the ratio of boys to girls is higher than this, and in the older groups it is lower. In fact, the ratio begins to decrease when mothers are only twenty-four years old. There is a similar decrease in the ratio of boys to girls born to fathers of advancing years. The reasons for these changes are not altogether understood, but they are compatible with the observation that the number of boy babies born during and after a war is always greater than at other times. The best explanation of this phenomenon lies in the fact that during wartime there are always more than the usual number of early marriages and very young mothers and fathers.

Specific studies of the connection between a mother's age and various aspects of her pregnancy have been carried out in many countries, with many thousands of women and children. One of the most closely studied correlations is that between maternal age and a disorder of the cartilaginous growth known as achondroplasia. Dr. Adrien Bleyer, who studied Down's syndrome, has also done work on this problem, and his conclusion was that many more children with this disorder are born to older mothers than to younger.

Mental deficiency, too, occurs more frequently in children born to older mothers. Dr. Neil A. Dayton, of Boston, reported that imbeciles, for example, appear among the children of women between fifteen and twenty-four only once in 1,300 births; in women between thirty-five and forty-nine, they appear once in 700 births, which is almost twice as often. Dr. Dayton ends his report on this subject by saying, "It offers a plea for the completion of childbearing before the mother is thirty-five years of age."

Stillborn children were found, by Dr. J. H. Edwards in a large-scale study in Scotland, to be three times as frequent among mothers between forty and forty-four as among mothers between twenty and twenty-four. Prematurity, particularly associated with the toxemias of pregnancy, is three times as frequent in women forty-four or over as in younger women. The main signs of the toxemias of pregnancy are hypertension, excessive protein in the blood, and in severe cases, convulsions and coma. The causes of these toxemias are still unknown, but obstetricians have made great strides recently in recognizing the early symptoms and in treating them successfully.

It is also believed, mainly because of work on lower animals, that the offspring of young women live longer than the offspring of older women. This is difficult to study in human beings, but it appears to be a reasonable extension of the studies of the general health of children of women of various ages. If the children of younger women are healthier than those of older women when they are babies and young children, it is quite possible that they may also be longer-lived.

On the other hand, the general health of a child depends to some extent on his position in the family, regardless of the age of his mother, and sometimes this effect outweighs the influence of his mother's age. For example, during prenatal life, the firstborn child in a family tends to be less healthy than the second and third children, in spite of the fact that the mother is younger during her first pregnancy, and presumably in good condition to nourish and protect her child. This was established in a study of over 600,000 births in England and Wales by Dr. T. McKeown, who charted the stillbirth rates of the children against both the age of the mothers and the number of their previous pregnancies. Dr. McKeown suggested that the reason for this might lie largely, if not entirely, in the fact that the placenta of firstborn children is usually smaller than that of the later-born in the same family. This would reduce the amount of nourishment and oxygen available to the child and affect his growth and development.

This same tendency was shown in the discovery in

Scotland by Sir Dugald Baird that a woman who already has several children is more likely to have a healthy child than a woman of the same age who is having her first child.

Another report that strenuously contradicts some of the findings on maternal age has been published by Dr. K. Fuchs and Dr. A. Peretz, who conclude from their work that there is no such thing as "too many pregnancies," at least so far as the health of either mothers or children is concerned. They based this conclusion on a study of more than 23,000 deliveries at the Rambam Government Hospital in Haifa, Israel, between 1949 and 1958. Of these women, over 14,000 were mothers of at least two children, and over 1,500 had delivered seven or more times. The size of this study, and the inclusion of many women with very large families, makes it particularly important. None of the women with seven or more children died, and the mortality rate of their children was no larger than that of the children of women with smaller families. And yet, a woman with seven or more children would normally be considerably older than a woman with only two. How is it that the mother's age does not appear to have affected these children? This is obviously a question that requires more study.

Although the firstborn child weighs the least of all the children in a family and runs the greatest risk of difficulties before his birth, his position changes as soon as he is born. No other child in his family will be so fortunate as he: firstborn children tend to grow and develop better after birth than their younger brothers and sisters. The reasons for this are not entirely understood, but some investigators have suggested that the economic position of most young parents is better at the birth of their first child than it is at the birth of the later ones. And he certainly receives more attention, particularly from his mother, than any of the others will have. As all mothers know, each additional child multiplies her duties and divides her attention.

Another factor that enters into the development of a child before birth, and that also may be connected with his mother's age, is the interval since the last pregnancy. Unusually prolonged intervals (exceeding eight years) seem to be connected with a higher rate of stillbirths than normal, but it is difficult to say whether this is due to the interval or to the fact that the mother has aged considerably in the meantime. An unusually brief interval (less than two years) also appears to increase the risk of a stillborn child.

This is a complex subject: not only the age of a mother and her whole rate of development and aging are involved, but also the interwoven strands that are the by-products of previous pregnancies. There is still much to study and to learn in this field, and investigators are adding to our knowledge with every passing year. Already, however, we know enough to be of use to a prospective mother: to warn very young women of the dangers of bearing children before their bodies are entirely ready, and to inform women over forty of the risks they are running.

VI—DRUGS

ONE OF THE GLORIES of modern medicine—the development and imaginative use of new drugs—presents one of the thorniest problems for prospective mothers and their doctors. These new drugs are making it possible to cure illnesses and conditions that in other days would have been fatal. They combat diseases, they lessen pain, they stimulate natural processes. They are the cause, more than any other single agent, of mankind's increasing life expectancy. And yet they present a clear danger to a child before birth; a prospective mother must approach the whole field of drugs with extreme caution.

When a pregnant woman takes a drug of any kind, her child will receive at least part of it, and sometimes that part will be larger than he can tolerate. Almost any substance can find its way from her bloodstream to her child's body. For that reason, one of the important jobs of both mother and physician is to make sure that no material that could harm the child is present in her blood.

But how can they know? They cannot. Where can they go to find out whether or not a specific drug will harm a child before birth? Nowhere. Some work has been done in this field, some observations have been made, and some studies have been carried out. However, compared with the vast number of drugs available, and the even vaster number of ways they can influence a child, the information we have is spotty and incomplete, and some of it is inconclusive.

The basic precaution for a prospective mother, then, is simple: she should not take any drugs of any kind, unless they are essential for her own health. In that case, she and her doctor should proceed with great care. This precaution includes such mild preparations as aspirin, and it even covers the anesthetics used by dentists during the extraction of teeth.

In addition, she should avoid drugs during all periods when she could be pregnant. There are many cases on record of damage having been done to babies before their mothers even knew they were pregnant.

A child before birth may be in danger from drugs for two reasons: his size, and the stage of his development. In these two important respects, a child before birth is an entirely different organism from his mother. Many drugs are prescribed according to the body weight of the patient, and a dose that is suitable for a woman who weighs 125 pounds most certainly is not the proper dose for a child who weighs three pounds.

Also, when a grown woman takes these drugs into her body, she changes them into other substances and absorbs them into her tissues in a form that is useful. A child cannot do this.

The liver is the organ in the human body where this breaking-down process takes place. In an adult, the liver forms a substance—an enzyme—that is capable of causing chemical changes in the drugs. They can then be assimilated into the bloodstream in their properly activated form. The liver of a child before birth is not capable of making these enzymes. Hence, the original drug remains unchanged and acts upon the child in a different way from that in which it acts upon the mother.

This subject is difficult to investigate in human beings, but work has been done on lower animals, and the development of the enzyme system of the liver has been clearly demonstrated in the laboratory. Mice and guinea pigs, for example, must be eight weeks old—that is, eight weeks *after birth*—before their livers are fully capable of breaking down various drugs.

A group of investigators from the U. S. Public Health Service have reported on an experiment that showed this. They used two analgesics for stilling pain, aminopyrine and phenacetin; a barbiturate, or sleep-inducing preparation, hexobarbital; a laxative, phenolphthalein; and a fever-reducing drug, monomethyl-4-aminoantipyrine. These drugs were not completely broken down in any of the animals under eight weeks old. The enzymes appeared for the first time in animals more than a week old. In the newborn animals, the drugs remained entirely unchanged. It follows, then, that in an animal before birth the drugs would also be unchanged.

It may seem a long step from the enzyme system of a baby mouse to that of a baby human being. Actually, in their basic principles, they are identical systems. There is every reason to suppose that a child's liver develops in much the same way as the liver of a mouse. And knowledge of comparative physiology and logic leads most authorities to the conclusion that in human beings the enzymes that break down harmful drugs do not begin to operate until after birth.

In that case, then, why are not *all* children whose mothers have taken drugs damaged by them? Surely, millions of women have been given drugs during pregnancy, and only a small proportion of their children have shown any ill effects. The answer to this question is that damage may have been done in many cases, damage which does not become apparent, if at all, until the passage of years. There are also remarkable constitutional differences in resistance.

An extreme example of this kind of resistance has been reported by Dr. A. Van Maarseveen. A twenty-two-year-old married woman attempted abortion during the fourth month of her pregnancy by taking 150 quinine sulfate pills of 50 milligrams each. When this was not effective, she took a small amount of rat poison paste, containing thallium sulfate. Still nothing. About a week later, the poison began to be noticeable in her own system: abdominal cramps, pains in her chest, and loss of hair. Her urine contained large amounts of thallium; in fact, it continued to show thallium for six months. Her child, a girl, appears not to have been affected in the slightest degree. She was born at term, seemingly normal, and never showed any evidences of poisoning.

On the other hand, groups of unmarried mothers show a higher proportion of malformed children than similar groups of women who are married. This difference can be accounted for in several ways, and it is a difficult subject to study, but some investigators have suggested that unmarried mothers commonly attempt abortion by using drugs, and when the attempts are unsuccessful, the drugs may, in some cases, be responsible for the malformations.

We must also recognize that not all damage to a child

can be spotted immediately after birth. And even when it appears later in the child's life, it may not be traced back to the original cause. A slightly weakened kidney, for example, showing up decades later, may actually have received the blow that weakened it before the individual was born.

In any case, we know already that human beings vary enormously in their reactions to all things. Some children will be seriously affected by an influence that will leave others entirely untouched. Under these circumstances, the only sensible course for a prospective mother is to take no chances.

A sleeping pill, for example, can be dangerous in any one of several ways. An experiment with mice of various ages, while not altogether applicable to human beings, suggests one of the ways a pregnant woman may affect her child when she takes a dose of hexobarbital. A small dose of this drug, a sleep-inducing preparation, was given to a group of mice. The dose was fixed by the body weight of each patient: ten milligrams for each kilogram of body weight. As a consequence of this dose, adult mice went to sleep and slept for five minutes. The same dose in proportion to their weight caused seven-day-old mice to sleep for 107 minutes. One-day-old mice slept for 360 minutes. What effect would this same dose, even adjusted for body weight, have had on mice that were still unborn? No one has inquired into that. But it certainly seems reasonable to suppose that if the dose were large enough to be of use to the mother, according to her weight, the effect on the not-yet-born baby might well be lethal.

Again, these results cannot be transferred entirely to human beings, but many of the same principles operate in all animals. At the present time, we have no way of finding out precisely how much of any substance in a mother's body succeeds in invading her child's system, but there is no real doubt that it is frequently too much for the child's own good. A long-term study of prematurely born children and children who were removed from their mothers by operations convinced Dr. Davenport Hooker of the University of Pittsburgh that barbiturates are among those drugs that clearly affect a child before birth. He studied a great many children whose mothers had been given barbiturates such as Amytal, monosodium isamyl, ethyl barbiturates, Nembutal, and pentobarbital. These children were completely anesthetized; their normal activity was totally eliminated. Other drugs, he found, were not quite so strong in their action, but even the least effective drugs substantially reduced the children's activity. These drugs were Avertin, ether, nitrous oxide, and morphine.

Even aspirin comes under the heading of drugs, when a pregnant woman is considering her child. Two aspirin tablets, of five grains each, is an ordinary dose for a woman trying to combat a headache. For her, this is a safe dose. But if she is pregnant, is it safe for her child?

The question almost answers itself. For her child, it is obviously a massive dose. Most experienced mothers would hesitate to give a two-year-old child, who has a fully operating enzyme system, any aspirin at all without first consulting a doctor. Then, if he agreed, she would give him no more than a quarter of the adult dose—that is, 1¼ grains. Yet when a pregnant woman takes ten grains of aspirin, she is subjecting her developing child to the possibility of a dose eight times as large, at a time in his life when he is probably unable to handle safely any at all.

There is only one solution to this problem: *avoid all drugs of all kinds at all times during pregnancy.*

Since aspirin is widely mistakenly regarded as a harmless drug, it is necessary to state here that investigation has shown that aspirin fed to mice and rats in their food from the sixth to the last day of pregnancy results in high death rates in the fetuses. In March, 1965, Dr. Daphne Trasler of McGill University reported the production of cleft lip in the fetuses of mice who had been given 15 to 25 milligrams of aspirin over a 24-hour period on days 8 and 9 or days 9 and 10 of gestation. In one strain of mice there were 9 percent of cleft lips and 19 percent of other malformations, while in another strain cleft lip occurred in 31 percent and other malformations in 24 percent.

A drug addict is at the other end of the scale from the woman whose problems are confined to the use of aspirin and the milder sedatives. Fortunately, drug addicts seldom become pregnant. Nevertheless, the problems of the child of a drug addict differ only in degree from the problems of the children of women who use milder drugs. Habit-forming narcotics—morphine and heroin, for example—have dramatic effects on a child, but the principles that govern those effects operate in the same way with milder drugs. A survey of some of the known facts about drug addiction is illuminating.

A woman who is a drug addict during pregnancy is almost certain to give birth to a child who, if he survives, shows the signs of drug addiction. His chances of achieving a normal birth are not good: the mortality and prematurity rates are high among the children of drug addicts. Furthermore, when they do survive, they are apt to be smaller than normal, and therefore more vulnerable to other stresses. The average weight of a normal baby is 7.5 pounds. The baby of a drug addict averages 6.7 pounds. Fortunately, addiction to narcotics tends to bring on an artificial menopause in women and impotence in men. For this reason, infant drug addicts are less numerous than might be expected from the number of adult drug addicts.

Treatment of a drug-addicted mother for her addiction during pregnancy is still a matter for debate. Some investigators believe that at any time before the seventh month of pregnancy, treatment has a chance of succeeding. Others place the last chance at the third month. Still others, pointing out the ease with which newborn children are cured of this kind of drug addiction, suggest that treatment be put off until after the birth of the child.

Curing the newborn child of an addiction to drugs is the only easy drug cure, according to one group of investigators, because it is the only addiction that is not rooted in psychological problems. The cure is accomplished by using paregoric or chlorpromazine.

When a nursing mother begins an addiction to a narcotic after the birth of her child, the narcotic will appear in her breast milk. Dr. P. M. Lichtenstein has reported the case of a mother who became addicted to heroin snuff immediately after labor; whenever the breast feeding of her child was delayed, or whenever she was unable to obtain heroin, the child showed clear withdrawal symptoms.

No abnormalities have been reported in the children of drug-addicted mothers. This does not mean that no damage is done to them; it means only that the studies which might throw some light on this have not been carried out.

Some investigators have suggested that drug addiction before birth might cause damage to the nervous system that it would be impossible to measure but that might express itself in personality difficulties later in life.

Eugene O'Neill, the greatest playwright America has produced, is a possible example of this. He was born of a mother who was a drug addict not only during her pregnancy but for a long period before it. O'Neill's character and personality were complex and difficult throughout his life; could this have been in any sense a result of his mother's drug addiction? And what connection, if any, existed between his stormy life and his magnificent plays?

There are no available answers to these questions, but they suggest a fertile field for investigators: the later lives of children of drug-addicted women. These studies would be extraordinarily difficult to organize and carry out, because drug addiction nearly always reflects deeper problems, and these would have to be taken into consideration. The grown children of women who were addicts during pregnancy would naturally be carrying combinations of such complex influences as might defy analysis by ordinary methods of investigation. In any case, future studies of this kind, if they are ever undertaken, will add to our knowledge of the effects of all drugs on children before birth.

A major barrier to our understanding of the effects of drugs is the problem of testing them. Testing drugs on adults is difficult and hazardous enough; testing them on children, before or after birth, is even more so.

A suggestion has been made by Dr. F. Clark Fraser, of McGill University, that the effects of drugs on children before birth might be studied among women who must have abortions. He also suggests that careful data be collected on the first few hundred or few thousand pregnant women who must use a new drug for their own health. These are good suggestions, and if they were carried out, we would have more information than we now have. But it would come in very slowly, and some of it would be obscured by other factors.

The most dramatic study and object lesson of recent years was presented to the world during the summer of 1962, when thalidomide was unmasked as the cause of thousands of deformities among newborn children in

West Germany. The popularity of this drug, and the discovery of its dangers, combine to give us a clear example of the threat that all such preparations hold for pregnant women and their children, and the difficulties that stand in the way of investigations.

Thalidomide was popular. Naturally; why not? It was a highly effective sleeping pill. Furthermore, the sleep was normal and deep, and it left behind no hangover to ruin the next day. Thalidomide was cheap; almost anyone could afford it. For a long time, it was available without prescription; anyone could buy it. It had been tested on animals and it had caused no damage to them; it seemed safe. No one could commit suicide with it. And it had still another advantage: it helped combat the nausea and vomiting of early pregnancy.

In West Germany, where thalidomide was first manufactured and sold under the name of Contergan, it was used almost everywhere. Doctors prescribed it for their patients at home, in hospitals, and in mental institutions. It was added as a sedative to other drugs and in that way became part of the treatment of grippe, neuralgia, asthma, and coughs. Hospitals gave it to children to quiet them before electroencephalograms, and to other patients when they were restless. Parents bought it in liquid form for their children and used it as a kind of baby-sitter.

Such an excellent drug obviously could not long remain the exclusive property of a small group of people in a single country. Before long, pharmaceutical companies in England, Canada, Australia, Holland, Portugal, Japan, and the United States were manufacturing it under various names—Distaval, Softenon, Kavadon, and Talimol. In all these countries except the United States, it was sold, and because of its fine qualities, it sold widely. Here, as every newspaper reader knows, it failed to pass the Food and Drug Administration's tests, and we were spared both the good night's sleep and the sad awakening.

A preliminary rumble, as of a distant storm, disturbed the clear atmosphere of the West German Never-Never Land, in April, 1961, when doctors there began seeing cases of a new form of polyneuritis. The symptoms were serious: tingling of the hands, disturbances involving both muscles and senses, and atrophy, or wasting away, of the tissues of the thumbs. Doctors soon recognized that the patients with these symptoms had been using thalidomide frequently and over a long period. They also learned that unless the patients stopped using the drug immediately and altogether, the polyneuritis was permanent. After this, Contergan was placed on prescription. Nevertheless, doctors still prescribed it freely, and it was still widely used in hospitals and homes.

Meantime, another malady was beginning to appear: phocomelia. This is a birth defect, extremely rare under ordinary circumstances, consisting of a malformation or even the absence of the bones of the arms and legs. Beginning in 1959, several cases of phocomelia appeared in maternity hospitals in West Germany. In 1960, the number of cases increased sharply. In 1961, there were ten times as many as in 1960. In other parts of the world it began to make its appearance—in other countries of continental Europe, in England, Australia, and Japan.

In September, 1961, Dr. H. R. Wiedeman reported a series of 33 of these children, describing their condition minutely and trying to arrive at some tentative generalizations about them. Other doctors in West Germany and elsewhere worked desperately on the problem. They sent out questionnaires to their patients and searched the records of all the mothers and children for some one item common to all of them. Some even wondered if we were not finally seeing a specific result of radioactive fallout.

Dr. W. Lenz, in Hamburg, was one of the most active of these investigators, and his work proved to be the most fruitful. His questionnaires, sent to patients and hospitals, inquired about all manner of things: X-ray exposures, hormones, foods, food preservatives, contraceptive measures, and tests for pregnancy. Among other items, Dr. Lenz's studies showed that 20 percent of the patients had taken Contergan. In spite of this small percentage, Dr. Lenz was suddenly struck with the suspicion that Contergan was the cause of the deformities. This was on November 8, 1961. He went back to the women and asked them specifically about the use of Contergan. Immediately the figure rose to 50 percent; many of the women said they had considered the drug too innocent to mention.

By November 15 Dr. Lenz had this information, and he notified the manufacturers of Contergan. On November 20 he reported to the Düsseldorf Pediatric Meeting that he suspected a single drug as the cause of the phocomelia. He did not name the drug, but during the evening a physician who had attended the meeting said to him, "Will you tell me confidentially: is the drug Contergan? I ask because we have such a child, and my wife took Contergan." During the next few days he received half a dozen letters asking the same question and making the same statement: "We have such a child, and my wife took Contergan."

By the end of November, doctors all over West Germany shared Dr. Lenz's suspicions, and the drug company had withdrawn the drug from the market. Furthermore, the Ministry of Health had issued a firm statement, which was carried on the front page of every newspaper and over every radio and television station, warning women not to take the drug.

During the same period, Dr. W. G. McBride, in Australia, was studying the backgrounds of six cases of phocomelia there. He came to the conclusion that the drug Distaval—the English and Australian name for thalidomide—had been taken by all the mothers early in their pregnancies and could be the cause of the deformities. He notified the drug company immediately; on November 27, the Australian company cabled the London office, and Distaval was taken off the market there.

By January, 1962, Dr. A. L. Spiers, in Scotland, proved to his own satisfaction that eight out of ten mothers whose children had been born with phocomelia had taken Distaval in early pregnancy.

Medical men in all the areas where phocomelia had appeared now began studies anew. Dr. Lenz undertook a particularly difficult work: to determine in how many patients who had given birth to infants with phocomelia he could obtain absolute proof that the mother had taken Contergan during pregnancy. By "absolute proof" he meant either a photostatic copy of a prescription or a hospital record showing the date and the amount of Contergan given. The difficulties were great: before April, 1961, the drug was sold without prescription, and many of the women who had taken it had no proof at all. Others, equally sure that they had taken it, were unable to remember exactly when. Some, however, had gone to the hospital for some minor operation before they knew they were pregnant and had received a sleeping tablet there. In many West German hospitals, sleeping tablets are given freely by the nurses and entered on the patient's record. Other women remembered taking a sleeping tablet because of some particular event that had made it necessary, and were able to track down the proof in that way. One woman's father-in-law had been murdered and she took Contergan for three nights after that. Another had watched a neighbor's farmhouse burn down. Another woman took Contergan and gave birth to a child with phocomelia, because the sedative her doctor had prescribed had not been available at the drugstore, and the prescription was stamped "Drug not in stock, Contergan given instead."

By the middle of March, 1962, Dr. Lenz had analyzed fifty cases in which he had interviewed parents and reviewed hospital records; forty-five of these women had taken Contergan between the thirtieth and fiftieth day of their pregnancies; five had taken it between the fiftieth and the sixtieth. Dr. Lenz concluded that the critical period—that is, the period during which thalidomide is dangerous—lasts for some two weeks, between the twenty-seventh and fortieth days of pregnancy.

The incidence of phocomelia in England is much lower than in West Germany, but at the present time there are some 350 cases in England and Wales, and over fifty in Scotland. Case have also been reported from Sweden, Belgium, Switzerland, Lebanon, Israel, Peru, Canada, Brazil, and Japan. Most of these cases were definitely connected with Contergan or Distaval or the Portuguese Softenon.

By the middle of July, 1962, the story of thalidomide became public in the United States, and no one who was reading the newspapers and magazines during the rest of that summer needs to be reminded of it. Dr. Helen Taussig, a distinguished physician of the Johns Hopkins University School of Medicine, had published a report in the *Journal of the American Medical Association* late in June, describing her investigation of the outbreak in West Germany and elsewhere; her report was the signal for the appearance of a large number of articles in general magazines all over the country.

Actually, we were never in serious danger from thalidomide. A few American women took thalidomide, not knowing what they were taking, because of our system of experimenting with drugs through physicians and their patients. A few more bought the drug in Europe. As a result, some American children were born with phocomelia. But thanks to our Food and Drug Administration's scrutiny of it, and Dr. Frances Kelsey's stubborn refusal to pass it until she was satisfied that it would cause no damage, we were spared the wave of defective babies that swept over West Germany.

There, 10,000 babies were born with phocomelia; 5,000 of them are still living. Sixteen hundred of these children will need artificial limbs, and many of them have internal malformations as well.

In spite of all the work that has been done on thalidomide, scientists are still not sure how it acts on a child before birth. The suggestion has been made that thalidomide resembles another substance—glutamic acid or a B vitamin—and that the rapidly dividing cells mistake it, as it were, for that other substance. In any event, the cells absorb it readily. Then, instead of nourishing the cells and helping them to grow, as another substance would do, thalidomide stops the division of certain cells, most frequently those of the arms and the legs. If thalidomide is present at the time when the schedule calls for the development of this part of the body, cell division fails to take place, and the child is born defective.

For those of us in the world who were spared, the story of thalidomide is a warning bell, clanging out an urgent warning: it can be dangerous to gobble down *any* pills indiscriminately and unthinkingly. We depend on our Food and Drug Administration and our doctors to protect us from harm from drugs, but this is not enough. Each of us must take personal responsibility, too, for ourselves and our children.

Doctors and drug manufacturers and even governments have taken heed of the lesson. A dozen European countries have forbidden the sale without prescription of drugs likely to be used by expectant mothers. In November, 1962, in Great Britain, an appetite-suppressing drug called Preludin was blamed for the birth of two deformed babies of one mother, the first in 1958, the second in 1960. Although this drug has never been suspected before, the

public concern over thalidomide is causing everyone to be
cautious, and the British Ministry of Health and the Poisons Board have both taken official notice of Preludin.
British doctors have also been warned about two drugs
that are widely used which have caused prenatal abnormalities in laboratory animals: Marzine, used for motion
sickness, and Imipramine, an antidepressant.

The West German pharmaceutical company that manufactures Preludin has warned women against taking
it during the first few months of pregnancy. The Italian
Ministry of Health has stopped the sale of Preludin and
thirteen other drugs. The Belgian government has placed
seven tranquilizers on the list of drugs obtainable only
with medical prescription. Manufacturers of Preludin in
Spain have published warnings to women not to use it
until further investigations of it are completed. Denmark,
Sweden, Norway, and The Netherlands have begun separate investigations of Preludin and other suspected drugs.
Denmark, Sweden, and Norway have also banned the
sale, without prescription, of Postafen, an antiseasickness
preparation.

In Washington, the Public Health Service and the Food
and Drug Administration have urged pregnant women to
take no drugs, particularly Preludin or Postafen, without
consulting their doctors. Both drugs are on sale in the
United States. The Food and Drug Administration has
also issued new regulations which require detailed reporting
to the government of clinical investigations of new drugs.

In the spring of 1963, a United States Senate committee held a hearing on drug-testing procedures. It was
clear from the testimony that much remains to be done
in this field, but it was equally clear that we are moving
in the direction of doing something. For example, Dr.
Norman Kretchmer, the head of the department of pedi-
atrics at the Stanford University School of Medicine, suggested that drugs intended for babies should always be
studied first for their effect on infant animals. This is an
eminently sensible suggestion. In view of the thalidomide
episode alone, without regard to all the other information
we have, what seems surprising is that this kind of testing has not already been universally adopted.

The most general and most impressive warning of all
was issued in the fall of 1962 by scientists gathered in

Ann Arbor, Michigan, at a seminar on birth defects. These specialists urged that pregnant women take no drugs, even the simplest. This includes, they said, even aspirin and the common antinausea preparations.

Their reason was straightforward: for all we know, any drug may be capable of harming a child before birth. This is not to say that a specific drug will be harmful. But it means that we cannot be sure it will *not* be harmful.

Even these scientists did not go quite far enough. They might well have warned all women who *might become* pregnant to avoid these same drugs. The most serious consequences of the action of drugs occur precisely in that period of early pregnancy when the major organs of the child are developing, which may be before a woman even knows she is pregnant. As Dr. Helen Taussig put it, when she was discussing the thalidomide situation, "Young women must learn that nothing is foolproof and new drugs should not be taken unless absolutely necessary." In fact, they should forgo drugs entirely during their entire childbearing period, except under the supervision of a doctor who is himself aware of the properties of the drugs he suggests to his patients.

The general rule for all prospective mothers to remember is this: all drugs are capable of crossing the placenta from the mother to the child. In Chapter II, we saw the mechanism of this process: the molecules, carried in the mother's bloodstream, move through the thin walls of the villi, into the placenta, through the umbilical cord, and from there into the circulation of the child.

The possible consequences of the passage of drugs from mother to child are numerous and varied, depending on the child, the drug, and the period during pregnancy when the mother takes the drug. In most cases of which there are records, the consequences are harmful to the child, although, as we shall see in the case of antibiotics, sometimes they may be beneficial.

The published reports on the effects of drugs may be misleading. In the first place, not all drugs have been studied. There are more than 9,000 drugs available in the United States, and more appear every month. The studies lag far behind the development of new drugs. This means that any survey of studies is of necessity full of holes. Secondly, most of the studies that have been made are of only a few individuals, and the conclusions are not necessarily final. Also, the studies of human subjects consist entirely of observations rather than experiments, and observations of human beings are notoriously chancy.

For example, some of the drugs that are used in the treatment of diabetes are now under suspicion as having caused damage to several children of diabetic women. In a group of twenty young women who had been treated with tolbutamide, the mortality rate of their children was higher than normal. Carbutamide is another drug of this kind that may be harmful, although the damage in this observation was limited to a single child. Chlorpropamide, too, is believed by Dr. G. D. Campbell to be responsible for the deaths of some of the children of another group of twenty-two diabetic women. None of these correlations can be proved, and in fact, many diabetic women who have been treated with all these and other drugs have had normal, healthy children, but suspicion of these particular drugs has been aroused. They should not be used for treatment of a pregnant woman unless nothing else is available.

Insulin shock therapy, used in the treatment of emotionally disturbed women, may also be harmful to a child. Dr. D. E. Sobel has reported five cases from his own experience and twelve from the literature, a total of seventeen women, whose children showed a high mortality and malformation rate. This is a small sample and is certainly not conclusive, but the high rates have some meaning, and until they are more closely identified, insulin shock therapy must be under suspicion.

Cortisone treatment may also be dangerous to a child. It may produce an imbalance in the carbohydrate metabolism, which is capable of setting off a series of difficulties involving the heart, the liver, the bloodstream, and the size of the body. The first few hours of the life of such a child are critical. If he survives those, his symptoms may disappear two or three weeks after birth. The doctor of a woman who must have treatment with cortisone during pregnancy must be particularly careful to keep her carbohydrate metabolism under control.

Quinine, too, is a risky drug. Medical men have known for many years that it may produce deafness in a patient who is being treated for malaria. We now know that children before birth are equally susceptible to this deafness. Because of its danger to the child, quinine is no longer used, as it once was, to induce labor. Doses of 600 milligrams a day are dangerous and should be avoided. Quinine water is probably harmless, since the quantity of quinine hydrochloride in a ten-ounce bottle is about twenty-five milligrams, and a woman would have to drink some twenty bottles of quinine water daily before she would absorb an amount large enough to be dangerous.

Antithyroid drugs, such as thiouracil and thiourea, sometimes cause a goiter in a child before birth. In the majority of cases, this soon returns to normal, but occasionally hypothyroidism develops and may persist for months. Quadrinal, a preparation containing potassium iodide, was responsible, in the opinion of Dr. M. P. Galina and his colleagues, for the death of two newborn children who were born with large goiters. A popular asthma remedy, Felsol, also appears to be capable of producing goiter in a child. There are three cases on record of adults whose goiters seemed to come from this preparation, and one infant whose mother had taken Felsol during her pregnancy was born with a goiter.

Dicumarol, an anticoagulant drug for treating blood clots, is believed to have caused the deaths of several newborn children to whose mothers it was given. There is another anticoagulant drug, ethyl biscoumacetate, which appears to be safer than Dicumarol for children.

Chloroquine, a drug sometimes prescribed for rheumatoid arthritis, occasionally causes new hair to grow in white. In one case where this happened, the patient was four months pregnant when she started treatment. When her child was born, it was clear that the drug had reached him and affected him exactly as it had affected his mother: the tips of his hair were dark, but all the rest was white.

Even vitamin K, normally a beneficial substance, can cause trouble. Some physicians leave a routine order for doses of vitamin K to be given every twelve hours during labor. However, it has been shown that in large doses this vitamin may produce a condition that is damaging to the brain. Dr. J. F. Lucey and Dr. R. G. Dolan point out that premature children are particularly susceptible to this condition, and if a woman who is about to have a premature child also has prolonged labor, the amounts of vitamin K can build up and become dangerous.

ACTH has been suspected of causing cleft palate in a few children whose mothers were treated with it during pregnancy. Cleft palate appears in one out of 2,500 children in the general population. Dr. A. M. Bongiovanni, in a review of the medical literature, found two children with cleft palates among 260 children whose mothers had been treated with steroids. This is not many, but these two children were among only ten children whose mothers were given unusually large doses during the first month of pregnancy. This seems to point to a connection between ACTH and the formation of the palate.

Sometimes, when a pregnancy is threatened with miscarriage, a doctor will prescribe treatment with progesterone, to stabilize the pregnancy. This is a hormone, normally secreted by the placenta, and the treatment is effective as far as the pregnancy itself is concerned. Recently, however, investigators have found that when progesterone is introduced into the system at the critical period of sexual development, it may cause the external genitalia of a girl child to become masculinized. There are several safe oral preparations now available for threatened abortion or miscarriage, but they should not be used until everything else has been considered.

Normal sexual development in a child before birth comes about as a result of the action of several hormones on the tissues very early in pregnancy. The sex of the child is determined at the moment of conception, but until he is five or six weeks old, he shows no genital traits of either sex. A pair of sex glands, the gonads, are present, and these will soon develop into either ovaries or testes. There is also a double set of ducts, one male and one female.

With normal development the sex organs begin to appear at the end of the sixth week. The appropriate duct system develops, and the other regresses completely. The various organs form, grow, and develop. By the end of the eighth week, the sex of the child is recognizable from its external organs.

The critical period of sexual development, therefore, lies between the sixth and ninth weeks of pregnancy. But there is further gradual development until after the fourth

month, and if at any time during this entire period the
mother's system is flooded with extra hormones, or syn-
thetic hormonelike drugs, they may enter the child's system, where they are capable of supplanting the existing
hormones, or counteracting them, or perhaps even stimu-
lating the child's own glands to abnormal activity. Any
of these responses may cause a girl baby to become
masculinized.

It must be emphasized here that very few women who
have taken progesterone have borne children whose sexual development was unbalanced. Nevertheless, experiments with lower animals, and observations of human beings, show a correlation between hormonelike drugs and
disturbances of sexual development.

Anticancer drugs are also dangerous to a child before
birth. These drugs are specifically designed to block the
undesirable cell divisions of cancer, and they would
therefore be particularly hazardous to an organism that is
developing by that same mechanism.

Damage to children before birth has also been traced
to the use of several other drugs. Among them are novo-
biocin, chlorpromazine, the sulfa drugs, and tetracycline.

Some antibiotics, on the other hand, may be highly
beneficial to a child before birth. In general, the antibiotics reach the child quickly and easily through the
placenta and give him powerful protection against diseases that may come to him from his mother. Aureomycin,
Terramycin, and penicillin are all useful drugs for this
purpose. They are not, however, without their hazards. A
cautionary note is sounded in this connection by Dr. J.
M. Sutherland, who suspects that chloramphenicol may
be associated with a high mortality rate in newborn children. Dr. E. Lenzi and Dr. F. Ancona echo this note, by
suggesting that the use of streptomycin be re-examined,
because of occasional cases of deafness in children whose
mothers were treated with it during pregnancy.

The greatest danger to a child before and during birth
from the milder drugs—those that are used as anesthetics
and analgesics—lies in the effect they have on the oxygen
content of the blood. As we have seen in Chapter IV,

anoxia, or oxygen starvation, is a possible consequence of the use of anesthetics, and this anoxia sometimes leads in turn to a degree of brain damage. Normally, if the pregnancy has been uncomplicated and the mother has had no anesthetics, the oxygen content of the baby's blood will begin to increase rapidly as soon as he is born. From about 50 percent saturation at birth, it will go up to 75 percent within the first two minutes of his life. However, if the mother has been given a moderate analgesic, such as Demerol, within an hour before delivery, the child's oxygen content will be lower at the start, and at two minutes after birth will have built up only to 49 percent. Normally, without analgesic, the oxygen content at six minutes will be 90 percent; with Demerol it will have climbed only to 72 percent by six minutes.

This is not necessarily a damaging situation. The hazards arise when complications appear. Then the presence of an oxygen-reducing agent may cause serious danger. For example, sometimes a woman is given an oxytocic drug to promote rapid labor. Occasionally, this drug causes prolonged contractions of the uterus, which, as we have seen, can cut down the oxygen supply to a point below the safety level. If this is combined with an anesthetic, there may be a sudden and dangerous fall in blood pressure. To counteract this, still another drug, a vasoconstrictor drug, is also sometimes given. This is a drug that produces a narrowing of the blood vessels. The total effect of all these conflicting drugs in a body that is already in labor is sometimes dangerous to both mother and child. In one group of 741 women, who had received either a vasoconstrictor drug, or an oxytocic drug, or both, investigated by Drs. G. N. Casady, D. C. Moore, and L. D. Bridenbaugh, thirty-four developed severe and disabling postpartum hypertension. One woman had a cerebral hemorrhage that caused a partial paralysis of the left side of her body and the loss of the ability to speak, to read, and to write. More than a year later, this woman was still seriously disabled. In such a case the child may also develop a cerebral hemorrhage.

Sedatives can also interfere with the breathing apparatus of the child. It is not possible to test the extent of this kind of damage on human beings, but considerable work on it has been done with animals. Information gained in this way is not altogether transferable—human babies and young rabbits are obviously not identical—but enzyme production is similar in both. And the experiments conducted at Johns Hopkins by Dr. F. F. Snyder are startling enough to give pause to all prospective mothers and their doctors. Dr. Snyder gave various analgesics to 130 rabbits during labor. He used Demerol, Nisentil, Methadon, and several morphine derivatives. Of the nearly 600 rabbits born, over 300 either were stillborn or died within twenty-four hours. This is over 50 percent, a shocking figure. In 95 percent of the dead rabbits, Dr. Snyder found that the lungs had been contaminated with blood, epithelial cells, and other debris. And in over half of these animals, large areas of the air passages were also blocked.

As a comparison, Dr. Snyder investigated 164 newborn rabbits whose mothers had received no drugs during labor. He used the total number born in thirty consecutive litters, in order to avoid any screening that might interfere with an honest comparison. In all thirty litters there were no deaths at all. The lungs showed only small amounts of red blood cells, and not one of them suffered from a blocked air passage.

Clearly, this is not typical of human experience. Many millions of women have been given sedatives during labor with no noticeable effect on their children, and many millions more will probably follow.

On the other hand, sometimes a child is affected before birth in a way that is not noticeable, that is almost not measurable, but that his mother would have avoided if she had known that the danger existed. A slight lessening of intellectual capacity seems to be one of these effects. As we saw in Chapter IV, when a large group of anoxic children was compared with a large group of children who were not anoxic at birth, the former tested slightly lower than the latter in general intelligence and vocabulary. The same effect was shown by Dr. S. G. Armitage, who investigated the effect of sedatives on the learning capacity of laboratory animals. He used rats, which are capable of considerable learning. Dr. Armitage gave various doses of barbiturates to mother rats during labor, and then later tested the young rats. He found that the rats whose mothers had received drugs during labor were definitely less able to learn and to solve problems than rats whose mothers had had no drugs.

Several other studies with animals come to the same general conclusions about sedatives during labor. After working with guinea pigs in this connection, Dr. R. F. Becker and Dr. W. Donnell wrote: "The warning to expectant mothers is obvious: it is not wise to insist on being 'snowed under' at delivery. A little pain at childbirth, mollified by light sedation, may prevent extreme sorrow later."

Doctors did not always believe this. In fact, this information is relatively new to everyone. Since the birth of medicine, physicians have devoted their lives to the relief of human pain and the saving of human life. Both of these objectives appeared to be served when anesthetics were first used in childbirth: the mother was relieved of the pains of labor, and the doctor, as a consequence, found that he could work better and more safely. Sir James Young Simpson, the Scottish obstetrician who first used ether and then chloroform in 1847, wrote that he believed that anesthesia would remove "the perils as well as the pains of labor."

Doctors and laymen alike approved the use of anesthetics, and in medically advanced countries it became commonplace to use heavy sedation during labor. Many American women who had their children during the second quarter of the twentieth century went through the birth process each time entirely unconscious.

During this period, Dr. Joseph DeLee wrote in his influential book, *Textbook of Obstetrics,* published in 1933, that "the patient must be willing to pay the price of occasional loss or injury to the child as the cost of her relief from suffering." Dr. DeLee was guilty here of believing that "the patient," who is after all an individual woman pregnant with a single child, is able to take the statistical view that he so lightly assumed. "Occasional loss or injury" is exactly the price an individual woman is *not* willing to pay, if there is any chance that the loss or injury may be that of her own child.

She can ensure the best possible chance for her child by arranging with her doctor for a regional instead of a general anesthetic, and by taking as little of it as possible.

Any survey such as this, which attempts to report on the effects of drugs on human babies before they are born, is bound to be only a progress report. The work, in laboratories and hospitals, that establishes the facts in this field is spotty and haphazard. The conclusions are published piecemeal in journals all over the world. Furthermore, the field is enormous and growing with every passing week, and the most devoted and skilled investigators can work on only one corner of it at a time.

On the other hand, even the most random collection of studies falls into a pattern. The conclusions differ in their details, but a strong family resemblance runs through all of them. In the small group of investigations reported in this chapter, a prospective mother can find a single fact emerging. The lesson is clear: drugs are medicines, to be prescribed by doctors and taken by patients only in the treatment of disease and disorders and conditions that cannot be eased by sensible living. Used in this way, drugs are a blessing. Used as a substitute for proper nutrition, sufficient rest, or healthful practices in general, drugs are a hazard to health and even to life.

Furthermore, young children are more vulnerable than adults to the dangers inherent in drugs, and more vulnerable still are children before birth. And the most vulnerable of all human beings are children in the first three months of their prenatal lives.

For these reasons, a pregnant woman or one who may become pregnant should never under any circumstances take any drug, however mild, unless it is absolutely necessary for her own health.

As Sir Robert Hutchinson writes in a "Physician's Prayer:"

From the inability to leave well alone; from excess zeal for the new and contempt for the old; from putting knowledge before wisdom and cleverness before common sense; from making the cure of disease more grievous than its endurance to spontaneous resolution; May the Good Lord deliver us.

VII—TOBACCO SMOKE AND OTHER POISONS

A CUSTOM LOATHSOME TO THE EYE, hateful to the nose, harmful to the brain, dangerous to the lungs. . . ." Thus wrote King James I of England, three hundred years ago, in *A Counterblast to Tobacco*.

There is no doubt that smoking is physically loathsome to the eye and hateful to the nose, as well as much else. There is also very little doubt that when King James said it was harmful to the brain and dangerous to the lungs, he was right. More than three centuries after he mounted his attack, most scientific investigators have reached the conclusion that smoking is harmful to human beings. A few diehards still contest this, but mostly on the grounds that it is not yet proved beyond all question. How they can continue to do so after the publication of the Surgeon-General's report on *Smoking and Health,* and the reports of the Medical Research Council and of the Royal College of Physicians in England is difficult to understand. There can, however, be no question at all that smoking is harmful to a child before birth.

The implication is clear for a pregnant woman who is also a smoker: stop smoking.

Is this possible? Many women who smoke heavily and have been smoking for years will say "No." They have tried to stop, perhaps, and failed. Some have tried over and over again and failed each time.

"Anyway, I don't inhale," one woman protests. "How can my smoking hurt my child, if I don't even inhale?"

Inhaling has very little to do with it. Tobacco smoke

is absorbed into the bloodstream through the mucous membranes of the mouth and throat, as well as through the lungs. When she takes tobacco smoke into her mouth, a pregnant woman is speeding potentially harmful materials on their way to the placenta, and through it to her child's bloodstream. Nicotine, of course, is present in tobacco smoke, and nicotine is known to act on the nerve cells and the respiratory center. In addition, there are carbon monoxide, methyl alcohol, carbonic acid, alkali, pyroline, hydrocyanic acid, arsenic, furfural, pyridine, collidine, and various tar products, all in some degree poisonous to the tissues of the human body.

In general, the dangers of smoking can hardly be news to the American public. One of the most widely publicized debates in recent years has been whether or not smoking can cause lung cancer. Evidence has been presented that shows the two are intimately connected, and although there is sincere debate on this, many scientists believe that the one causes the other. As King James said, 350 years ago: ". . . dangerous to the lungs. . . ."

Has this affected the consumption of cigarettes? Of course not. In addition to being harmful and dangerous, smoking is an extraordinarily pleasant habit. And smokers, who have the human ability to rationalize their desires, can tell themselves that "It hasn't really been proved," or that "It won't happen to me," or "It's worth it," or even flippantly, "If I give up smoking, I'll eat too much, and everybody knows *that's* not good for you." The rationalizations are many and ingenious and constitute a clear indication that many adults have no real intention of ever giving up smoking, whatever the facts are. It is a waste of time to present them again.

A fact that is new to the American public, however, and more important to some individuals than the remote possibility of lung cancer, is this: smoking by a pregnant woman can definitely harm her child before he is born. And this may be an entirely different story.

A prospective mother who is genuinely interested in the welfare of her child does not wait for "proof"; the bare possibility that she may be hurting her child is enough to give her a reason to stop smoking. Nor is she likely to say, "It's worth it," because in this case the price for a selfish pleasure is exorbitant: possible damage to her child. For this reason, there is some justification in presenting the established facts on smoking by pregnant women and its effect on their children. The purpose of this chapter is to present these facts, in the hope that at least some prospective mothers will conquer their habit long enough to protect their children from this particular danger.

Oxygen starvation is again the chief threat. This is the most dangerous single aspect of smoking from the point of view of a child before birth, and a prospective mother who is a heavy smoker is referred back to Chapter IV for an account of the various effects of oxygen starvation on a developing child. Gases of all kinds—and tobacco smoke is, of course, a gas—rob the blood of oxygen. This is the same mechanism that operates when anesthetics enter the system. The mother's blood takes up the other gases at the expense of the oxygen. They enter the child's system, again in place of some of the oxygen. But his body is growing and developing at a rapid rate, and he needs oxygen desperately in order to carry on these processes. He needs it in every cell in his body, and if it is not forthcoming, his entire body can be affected by the lack.

Obviously, he will not stop growing and developing altogether, even if his oxygen supply is considerably diminished. Millions of normal, healthy children have been born to women who were heavy smokers. And women who have had several healthy children in spite of having smoked a pack of cigarettes every day during all their pregnancies will perhaps wonder if the danger to a child is really sufficient to warrant the discomfort of giving up smoking.

Certainly it is, although every woman must convince herself of it in her own way. It is quite possible that some women would go through considerable anguish in the process of giving up smoking, and this would not be good for the child, either. On the other hand, this kind of anguish is not difficult to manufacture, and a prospective mother must examine her reactions closely and carefully and honestly if she finds herself deciding that she cannot possibly give up smoking while she is pregnant.

The probability is that her healthy children would have been still healthier if she had not smoked during her previous pregnancies. And they would probably have been bigger. They might well have been a little brighter. And if any of them was premature, or particularly small, or was born earlier than expected, or had too many illnesses during his infancy, the reason may have been connected, directly or indirectly, with his mother's smoking.

The most dramatic correlation between smoking and the development of children before birth is to be found among premature children. There is no question in the minds of investigators that smoking is closely connected with prematurity. And since premature children are less able to handle the stresses of life than children who are born at term, smoking is basically the cause of numerous other results: some illnesses, weaknesses of various organs that may later cause trouble, lowered intellectual capacity, even, in extreme cases, death.

This correlation has been found in several studies of unusually large numbers of women. In San Bernardino, Cailfornia, for example, 7,500 women were the subjects of a study by Dr. W. J. Simpson. She correlated the smok-

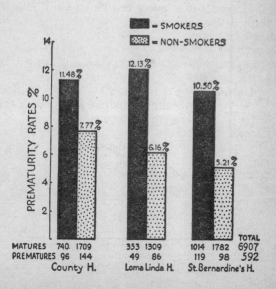

FIGURE 4. Showing percentage of premature babies born to smoking and nonsmoking mothers (Simpson).

ing habits of these women with the histories of their deliveries, paying particular attention to prematurity. The
patients were drawn from the public San Bernardino
County Hospital and from two private hospitals. In the
private hospitals, the women who did not smoke at all had
only half as many premature children as the women who
smoked. In the public hospital, the nonsmokers had two-
thirds as many premature babies as the smokers. (See
Figure 4.)

In addition, Dr. Simpson found that the women who
smoked moderately had fewer premature children than the
women who smoked heavily. Among those who smoked
between one and five cigarettes a day, the prematurity
rate was only 7 percent. Among those who smoked between sixteen and twenty, the rate was over 13 percent.
From there, the figures mount steeply: among mothers who
smoked between twenty-one and thirty cigarettes every
day, 24 percent of the children were premature, and
among those who smoked over thirty-one cigarettes daily,

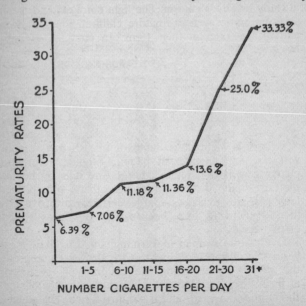

FIGURE 5. Relation of prematurity rate to number of cigarettes
smoked per day (Simpson).

33 percent, or one-third of all their children, were premature. (See Figure 5.)

Dr. Simpson also plotted the prematurity rates of heavy smokers, light smokers, and nonsmokers against the ages of the mothers. At every age, nonsmokers had the fewest premature children, and light smokers had fewer than heavy smokers. As we might expect from studies of women of various ages discussed in Chapter V, the mothers below the age of eighteen had a higher prematurity rate, whether or not they smoked, than women in their twenties. This is true of the general population, and the reason lies in the immaturity of the reproductive system of many girls of this age. Normally, as we saw in those studies, women in the age range twenty-three to twenty-nine have fully matured reproductive systems and have the fewest premature children. An important exception to this, however, is the group of women of this age who are heavy smokers. This is the only age group in which the prematurity rate *goes up* after the age of eighteen. The light smokers and the nonsmokers have fewer premature children in their twenties and thirties than in their teens. In contrast, among the heavy smokers the prematurity rate is high when they start having children, and it climbs still higher. At the age range thirty to thirty-five, for reasons that we do not understand, the prematurity rate drops for all groups. After that, the rate for heavy smokers climbs again until it is twice that of nonsmokers. All this information is shown in Figure 6.

When the meticulous detail of this study is combined with the large number of patients studied, it presents a picture that is difficult to ignore. And it is by no means the only study of its kind; several others, equally meticulous and, although not so large, certainly inclusive enough to be convincing, have been carried out in several parts of the world.

Over 2,000 women in Birmingham, England, were investigated during 1959 by Dr. C. R. Lowe. He approached the subject by weighing the children of these women at birth, then correlating the weights with whether or not the mothers smoked, and if so, how much. His conclusion was that the children of women who did not smoke were heavier by an average of nearly half a pound than the children of women who smoked. He also found, as Dr. Simpson in San Bernardino had previously discovered,

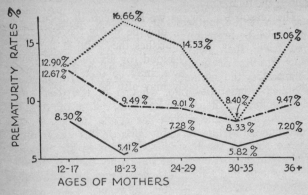

	Mat.	Prem.	Mat.	Prem.	Mat.	Prem.	Mat.	Prem.	Mat.	Prem.
NON-SMOKERS	298	27	1780	102	1375	108	922	58	425	33
LIGHT SMOKERS	62	9	515	54	434	43	229	21	86	9
HEAVY SMOKERS	27	4	255	51	283	48	154	14	62	11

FIGURE 6. Showing the relation of nonsmoking, light smoking, and heavy smoking among mothers to the births of mature and premature babies, and the relation of smoking habits to percentage of prematurity by mother's age (Simpson).

that the fewer the cigarettes a mother smoked per day, the heavier her child was likely to be. Dr. Lowe's study drew the line at ten cigarettes a day: the children of women who smoked less were heavier than those of women who smoked more.

Dr. Lowe also found that the boys of regular smokers were appreciably lighter in weight than the girls of non-smokers. Since newborn boys are on the average heavier than newborn girls, this reversal of the usual sex difference in weight is particularly interesting.

Furthermore, fewer of the children of nonsmokers died at birth, or were malformed, than of the children of smokers. This does not necessarily mean that smoking, in itself, is responsible for deaths and malformations; this is not one of Dr. Lowe's contentions. The figures do suggest, however, that where other conditions have weakened a child—poor nutrition, possibly, or an episode in early pregnancy involving anoxia—steady heavy smoking can deliver the final blow.

The reason for this, as we have seen in discussions of the effects of anesthetics, may be partly the oxygen content of the blood that nourishes the child. If it is low in oxygen, it is low-quality blood for his purposes. Another reason, suggests Dr. Lowe, may be that the quantity of blood in the child's body may also be diminished. There is no doubt that the substances in tobacco smoke tend to constrict the blood vessels of an adult; there is no reason to suppose they would have any different effect on a child. Thus a smaller amount of blood—as well as a lower quality of blood—would be available to his tissues. Smaller amount of blood: less food and oxygen. And this could easily retard his growth, physical and mental. When this constriction is repeated ten times a day, or fifteen, or twenty, or more—and each time for the number of minutes it takes to smoke a cigarette—the amount of nourishment and oxygen that is withheld from the child's tissues is enough so that the loss can be measured on scales by the time he is born. And there is doubtless some degree of loss—except that it cannot be so easily measured—in more specific functions and structures.

Dr. Lowe also wonders if smoking by pregnant women has not negated much of the progress in obstetrics in recent years. These advances should properly have led to a corresponding decline in the number of premature babies. Instead, it appears to have had no effect on the prematurity rate at all; this has remained fairly constant, despite the best efforts of physicians and hospital staffs to lower it. Dr. Lowe's explanation of this puzzle is interesting: during this same period, the number of women who smoke has risen steadily. As a result, the number of children of low birth weight has been determined, not by the better care they get from their doctors, but by the poorer care they get from their mothers. A premature child is defined as a newborn infant who weighs less than 5½ pounds. "If the effect of smoking on birth weight is as considerable as this study indicates," Dr. Lowe writes, "it is evident that it must make a substantial contribution to the frequency of 'premature births,' defined according to weight." In other words, the two forces cancel each other out: improved obstetrical care, which is aimed at providing the best possible nourishment for a developing child, has been foiled in many individual cases by increased smoking. This has been frequent enough so that the proportion of premature children has remained constant during a period when doctors know more than they have ever known about preventing premature births.

If Dr. Lowe's conjectures are valid, then, a mother who smokes heavily during her pregnancy is actually working against her obstetrician.

This speculation has received strong support from a study of over 2,500 Negro women in Baltimore City Hospital. Those women who acknowledged smoking a pack or more of cigarettes a day had twice as many premature children as did the nonsmokers. The figures were approximately 23 percent for heavy smokers, as compared with approximately 11 percent for the nonsmokers. The proportion of premature children of all smokers—heavy, moderate, and light—was 18 percent.

These investigators—Dr. T. M. Frazier and his coworkers—confirmed the previous studies, too, as to the weights and death rates of the children of smokers, compared with those rates among the children of nonsmokers. In every category, the latter fared better, sometimes dramatically better, than the former. For example, when Dr. Frazier investigated the death rates of children before birth, he found that more than 15 out of every 1,000 children of smokers died before birth, whereas among nonsmokers, less than half that number died—approximately 6 out of every 1,000. Even this is a relatively high figure; this is undoubtedly a case of other factors being at work among mothers and children. Negro women having their children in Baltimore City Hospital are quite likely to be among the city's most impoverished citizens, and statistics on all of their activities and conditions will certainly reflect that background.

This same variation in the effects of smoking according to the variation in social and economic backgrounds was brought out in a study in Aberdeen, Scotland. The result of this investigation, which covered over 2,000 women, was published in 1962 by Dr. A. Herriot and his colleagues, Drs. W. Z. Billewicz and F. E. Hytten. The women were divided into three groups: the wives of professional men and managers, the wives of skilled workers, and the wives of semiskilled and unskilled workers.

In all of these groups, the children of nonsmokers were in better condition at birth than the children of smokers. They weighed more, and there were fewer premature births. The difference in the professional and managerial class was slight. Among the other groups, the difference was similar to that found in the studies in San Bernardino and Baltimore: about twice as many premature children to smoking mothers as to nonsmoking mothers.

The Aberdeen study also disclosed that smoking mothers appear to have fewer male children born alive than nonsmoking mothers. By pooling data from published reports, Drs. J. F. Fraumeni, Jr., and F. E. Lundin, Jr., of the National Institutes of Health at Bethesda, in 1964 confirmed the Aberdeen findings, and themselves found that the sex ratio in smoking women was reversed, that is, smoking mothers gave birth to more female children than male children. The general reason for this is that male children are more vulnerable before birth to all forms of disturbing stimuli, including tobacco smoke.

A condition called "fetal distress" was also found more frequently among the children of smokers than of nonsmokers. This is characterized by an abnormal heart rate, either exceptionally high or exceptionally low. As long ago as 1935 and 1938 investigators discovered that a single cigarette smoked by a pregnant woman can cause her child's heartbeat to change suddenly in either direction. Dr. Lester W. Sontag and his colleagues measured changes as great as thirty-nine beats more, or seventeen beats less, per minute. The significance of this is not clear, but it gives a prospective mother something to think about as she lights her cigarette.

Dr. Herriot and his colleagues concluded with a statement that from a normally cautious source is particularly strong:

Until we know more about the effects of smoking in pregnancy, it seems sensible to warn patients that it may be harmful to the baby. In patients who have a history of poor reproductive performance, particularly where this has been associated with unsatisfactory fetal growth, one may be justified in forbidding cigarettes.

The largest single study on this subject is the nation-wide long-range U. S. National Institute of Health Perinatal Research Project, begun in 1958 and still continuing. The study has already compiled data on more than 30,000 pregnant women and 24,000 children at fifteen participating medical centers. In July, 1962, some early findings were announced, and it is interesting but not unexpected to see that they corroborate the many other studies of the effects of smoking: premature births occur more frequently among the children of women who smoke than among those of women who do not smoke, and the more cigarettes a woman smokes, the smaller her child is likely to be.

Abortion has also been found by several independent investigators to be significantly more frequent in smoking than in nonsmoking mothers. An abortion is the birth of a baby prior to its capacity to maintain life, that is, before about 20 to 28 weeks of gestation.

In 1949 Dr. P. Bernhard reported the effects of smoking in 112 pregnant German women who were smokers compared with 1,381 nonsmoking pregnant women. The abortion rate was 22.5 percent in the smokers as compared with 7.4 percent in the nonsmokers. Prematurity, still-births, vomiting, difficult labors, edema, and pre-eclampsia were also more frequent among the smoking mothers.

Late in 1962 Dr. H. J. Heron reported on a detailed study of 58 smoking mothers and their infants compared with a control group of nonsmoking mothers and their infants. Special attention was paid to those mothers who continued to smoke during the first stage of labor, and to their infants. Among the smoking mothers, the carbon monoxide level of the blood of both infants and mothers was twice as high as that of the nonsmoking mothers and their infants. With the exception of three infants, all the babies of nonsmoking mothers cried lustily within half a minute of birth, whereas 81 percent of the babies of smoking mothers were delayed in regular respiration and crying for as much as three minutes. The incidence of prematurity in smoking mothers, as compared with nonsmoking mothers, was three to one. The babies of smoking mothers weighed less than those of nonsmoking mothers.

Early in 1963 Dr. J. M. O'Lane reported on 1,031

pregnant white American woman, 566 of whom were non-smokers and 465 smokers. The infants of smoking mothers weighed about 1½ ounces less than those of nonsmokers, and they were shorter by half an inch; the prematurity rate was 11.8 percent compared with 5.1 percent in the nonsmokers; the abortion rate was 12.6 percent compared with 9.9 in the nonsmokers. Interestingly enough, the smoking mothers had 112 abortions prior to the current pregnancy compared with only 91 in the nonsmoking mothers.

Also in 1963 Dr. J. R. Zabriskie reported on 2,000 women studied in Hawaii. There were 957 pregnant smokers and 1,043 pregnant nonsmokers. The infants of the smoking mothers weighed 8 ounces less than the infants of the nonsmoking mothers; the prematurity rate was 9.9 percent in the smokers compared with 3.8 percent in the nonsmokers, and the abortion rate was 12.6 percent in the smokers as compared with 8.8 percent in the nonsmokers. These figures, it will be observed, are virtually identical with O'Lane's published a few months earlier in the same year.

In February, 1964, Dr. Y. Yerushalmy reported the results of an investigation of 5,381 white and 1,149 Negro, single live births where the smoking status of the mothers was known. It was found that the prematurity rate was substantially higher among smoking than among nonsmoking mothers. A puzzling finding that emerged from this study was that the low-birth-weight mortality rate was substantially lower for infants of smoking mothers than for those of nonsmoking mothers. As Dr. Yerushalmy suggests, it is unlikely that smoking as such has any causal relation to the observed difference. Smoking women may represent a group whose reproductive experience would have duplicated the observed pattern whether they smoked or not. Only further research will give us the answer to this interesting question.

Other approaches to this same subject have been made by other investigators, who have studied the effect of some of the specific constituents of tobacco smoke on a developing child. These studies show us clearly the mechanisms by which damage can be done. Some of the most important of these have to do with carbon monoxide.

Carbon monoxide is one of the most dangerous of gases to a human being of any age; this has been demonstrated many times with both experimental animals and human subjects. It combines with the hemoglobin of the red blood cells to form carbon dioxide. This means that when carbon monoxide is present in the blood, the amount of oxygen is certain to be reduced by that amount. If the oxygen is reduced substantially, all the undesirable results of oxygen starvation follow. Dr. A. L. Barach and his co-workers have made studies showing that a nonsmoker carries 12 percent more oxygen in his blood than a smoker, and Drs. W. Haddon, R. E. L. Nesbitt, and R. Garcia, in studying pregnant women specifically, found that smoking produced a significant concentration of carbon monoxide in their blood. Since carbon monoxide is known to have a detrimental effect on the functions of cells, this can be a particularly dangerous aspect of smoking.

Carbon monoxide in the blood is also believed to slow down several important chemical changes that occur during a child's prenatal life. In the tightly organized schedule that governs his development, an unscheduled change in any single aspect can have far-reaching effects on the child's entire body and mind.

Nicotine, too, is thought to have an effect on the growing cells. Very little research has been done in this field with human beings, but an interesting experiment with fertilized eggs of sea urchins showed clearly that nicotine was responsible for a significant slowing down of cell division. It is not known at present whether this also applies to cell division in human beings, but until it has been proved that it does not apply, the sensible course would be to act as though it did.

Other gases, as well as tobacco smoke, are capable of causing difficulty to a child before birth. Carbon monoxide, of course, is encountered elsewhere, notably in the air of the densely populated, traffic-laden streets of our large cities. Industrial plants, too, are frequently guilty of belching into the air vast quantities of chemicals. When these are breathed into the lungs, they find their way into the bloodstream, drive out the oxygen, and in the case of a pregnant woman, eventually reach the circulatory system of her child. Here the damage done depends on the chemical itself, its concentration, and the stage of development of the child. Sometimes, obviously, the damage is so slight as to be unnoticeable. But it is certainly not good for any child to be nourished on this kind of diet—the less of it, the better for him.

Most inorganic poisons are also capable of reaching a child through his mother's body. Some of these substances are lead, mercury, arsenic, copper, phosphorus, bromide iodide, potassium chlorate, strontium, and yttrium. A careful mother goes to considerable lengths to keep from her young child toys and other objects painted with paints containing lead. She keeps well out of his reach all the household items—insect poisons, thermometers, medicines—that might harm him. It goes almost without saying that even greater care must be taken with her child before he is born.

Yet another hazard of cigarette smoking is radioactivity. In January, 1964, Drs. E. P. Radford, Jr., and V. R. Hunt of the Harvard School of Public Health announced the discovery of radioactive Polonium-210 in tobacco smoke. Ionizing radiation of this sort (see pp. 204–220) is known to be damaging to living tissues and productive of cancer. At the temperature of a burning cigarette, 600° to 800° centigrade, polonium is highly volatile and readily attaches to the smoke particles. These pass into the bloodstream, where the freed polonium will, among other things, pass across the placenta to the child. This latter has not yet been demonstrated because no one has undertaken this particular aspect of the investigation, but there cannot be the least doubt that such transplacental transmission of polonium occurs in smoking mothers. It is estimated that an individual smoking two packs of cigarettes a day would absorb about seven times as much radiation as that derived from background sources. Since polonium disintegrates quite slowly, its cumulative effect can be serious.

In August, 1964, Dr. Larry D. Samuels of the Harvard School of Public Health reported that Polonium-210 levels in the ovaries of women cigarette smokers were higher than in nonsmoking women. Dr. Samuels also reported that the semen of men who smoked two packs a day contained significantly higher levels of polonium than that of nonsmokers, and that pack-a-day smokers and pipe smokers had levels in between.

Unexpectedly, alcohol in the form of beverages, even in immoderate amounts, has no apparent effect on a child before birth. A mother who drinks heavily is not likely to be a good mother, but the reason is not that the alcohol itself harms her child, but that her everyday life is probably unhealthful, untidy, and unstable. Her nutrition is usually poor, and this has a direct influence on her child's development. But it can now be stated categorically, after hundreds of studies covering many years, that no matter how great the amounts of alcohol taken by the mother—or by the father, for that matter—neither the germ cells nor the development of the child will be affected.

Drunkenness, however, can be such a dramatic event, and the effect of habitual heavy drinking can be so tragic, that it is no wonder that men have believed down through the ages that alcohol must certainly affect the undeveloped, vulnerable child. The myth goes back at least as far as Vulcan, the blacksmith of the Roman gods, who was born lame, because, the Romans believed, his father, Jupiter, was drunk when he begot Vulcan. In ancient Carthage, bridal couples were forbidden to drink wine because Carthaginians believed that the result would be a defective child.

The facts, however, are quite otherwise. An amount of alcohol in the blood that would kill the mother is not enough even to irritate the tissues of the child. Part of the reason for this may be that alcohol is eliminated rapidly from the adult body; only very small quantities ever reach the child.

Many publications claim to show that alcoholics give birth to more idiots, more malformed children, and more retarded children that do sober individuals. This claim simply does not stand up to scientific examination. When studies are made of the children of intemperate parents, who themselves came from intemperate families, investigators find that there are no more feeble-minded, epileptic, or retarded children among them than among the children of temperate families. What differences there are between the two groups are a direct result of the mothers' inadequate nutrition and the families' poor environment, and not of the alcohol itself. A drinking mother harms her child before birth much less, it appears, than a smoking mother.

In view of all the studies of many thousands of women over a period of several years, there is no longer any room for doubt that tobacco smoke is one of the principal carriers of damage to children before birth. It is not so dramatic an agent, perhaps, as a drug calling attention to itself in all the popular magazines. On the other hand, it is more widespread than anything else in our society. The amount of damage in each individual child may seem to be small: a slight loss of health, a few ounces of weight, an amount of brain damage so small that no one will ever be able to prove that the brain was damaged at all, and only now and then a child who dies of his prematurity in spite of all that the doctors and the hospital can do to keep him alive. But when this kind and degree of damage is multiplied by the number of women who smoke, the loss to the rest of the world is staggering.

The loss to the child himself may turn out someday to be crucial, too. A prospective mother who is finding it difficult to give up smoking might consider this: her child's *entire life* is being decided in many of its aspects during the period she is carrying him. Her smoking or not smoking may determine more than whether or not he will be premature; in any case, the outward signs of prematurity are usually gone by the time a child is two or three years old, when he catches up with his more fortunate friends. But her smoking may also influence his life in other, less predictable ways. What young mother can look forward to the time when the child whose life is just beginning will be a middle-aged man? Yet that day will come, and when it comes is it not likely that he will need, for example, a strong, healthy heart? Suppose it has been weakened by a thoughtless mother before he was born? He may not survive beyond middle age because of it. And if his life should present him with severe problems, he will have particular need of a strong, healthy nervous system. Will he have one that is equal to his needs, or will he not?

The vast majority of children are entitled by their inheritance to healthy bodies and functioning minds. By the simple act of smoking a pack of cigarettes every day while she is pregnant, a mother can appreciably reduce her child's chances of receiving this inheritance.

Does she really want to do this?

VIII—NONINFECTIOUS DISORDERS

A CHILD DEVELOPS BEST, and with least help from the outside world, if his mother's body is functioning satisfactorily during her pregnancy. With that environment, all he needs is sufficient food of the right kind, and enough oxygen, and his growth and development will proceed smoothly from beginning to end.

But what if his mother is not in satisfactory health? Many pregnant women are not. By the time they are in their twenties, most women have gone through several illnesses, any one of which might have left behind a weakness. Some have been hurt in accidents. And during two dozen-odd years of life, a human being is exposed to all manner of bacteria and viruses. Any one of these factors may have affected a woman's body so that it functions less well than it once did.

Some women, too, were themselves born with slight deficiencies. A thyroid imbalance, for example, is one of the more common difficulties of this kind, and also one of the most easily treated. A thyroid gland that functions less effectively than it should is not necessarily a serious problem to a grown woman, and many women have gone through their lives with untreated thyroid deficiencies. They would have had more energy, perhaps, if the condition had been properly treated, but at no time were they seriously disabled by the deficiency.

A condition of this kind may well have had its start a full generation before, during a woman's own prenatal life. The malfunctioning of a thyroid gland can be passed on, not by inheritance—that is, not in the genes—but through the prenatal environment. A defective thyroid gland in a pregnant woman will secrete too much or too little of the hormone that this gland produces. This excess or deficiency will be felt in the body of the child and will affect in one way or another the development of the child's own thyroid gland. If this child is a girl, and if her thyroid deficiency is allowed to go untreated, she can pass some form of the disorder on to a child of hers in the same way. This is the way some of these chains are forged.

In the case of the malfunctioning of the thyroid, the chain is easy to break. A pregnant woman whose doctor knows of her deficiency—and it is readily spotted—can be treated so that her child will not be affected. Many other conditions can be similarly treated.

Conditions of this kind are known as noninfectious disorders. They are obviously not contagious; they affect a child before birth by way of the basic machinery of the mother's body: heart, lungs, glands, spleen, pancreas, liver, kidneys. These organs are working all through pregnancy in behalf of her child, and if the performance of any of them is less than good, the child is bound to be affected.

This chapter is concerned with some of these conditions, and with the various methods that have been worked out to remedy them.

Diabetes is one of the best-known examples of a functional disorder. The common form, diabetes mellitus, is caused by a deficiency of the pancreas. This gland normally secretes the hormone insulin, which enables the body to store and absorb sugar. When diabetes exists, the sugar of the food is not burnt by the tissues and converted into energy. A primary symptom of the disease is an excessive outpouring of urine with a high sugar content.

When diabetes in a prospective mother is recognized and treated, with diet and insulin and exercise, the danger to a child before birth can be considerably reduced. This is the conclusion of studies undertaken by Drs. E. R. Carrington, H. S. Reardon, and C. S. Shuman. The mother continues her familiar treatment, adjusting the amount of insulin and her diet as her needs and her child's needs change. Her physician helps her with this program, and together they maintain the health and well-being of the child.

After the child's birth, the physician keeps a sharp lookout for special problems. These may begin immediately, but if they are expected and prepared for, they can all be handled successfully.

One of these problems is concerned with the size of the child. Babies of diabetic mothers grow very rapidly in the womb and sometimes appear to be exceptionally healthy and well developed at birth, when in fact they are not. This can be a danger, because an inexperienced mother may not realize that her child is in reality rather frail and needs special attention. For example, he must be kept warm. This is standard treatment for all premature and small babies, but the child of a diabetic mother, because of his unusually large size, may not receive it unless his mother has been alerted.

The most common difficulty a child of a diabetic mother experiences is in his breathing apparatus. During the first twenty-four hours of life, he is more susceptible than most babies to the formation in his respiratory tract of a glasslike substance known as hyaline membrane. This is not really a true membrane but an exudate from the air cells which coats them over. This interferes with breathing, and sometimes even prevents the functioning of the lungs. The treatment for this condition is to clear the upper respiratory tract, use a tracheal catheter, and administer oxygen. Such a child should also be stimulated during his first twenty-four hours to cry and breathe deeply. This will tend to reduce the formation of exudate and allow air to enter the lungs.

The calcium content of a child's blood is another important matter. Even in the child of a healthy mother, this falls to a low level for two or three days after birth, and in the child of a diabetic mother this condition is sometimes aggravated, and may become dangerous. A low calcium content of the blood is sometimes associated with a disorder called tetany, which is characterized by muscular spasms. Investigators are not yet sure that the tetany is actually caused by the calcium deficiency, but the two sometimes occur together. In any case, the calcium content is another item to be checked and adjusted in the newborn child of a diabetic mother.

Another normal drop that is aggravated in such a child is in the blood-sugar level. The child has been dependent during his prenatal life on a particularly abundant sugar supply from his mother. When birth comes and he is separated from this, the normal drop is sometimes increased to a point where it can become dangerous. This condition is known as hypoglycemia, and it can be prevented or treated by feeding the child dextrose at intervals during the first few days after birth. In any case, the child of a diabetic mother should be examined for blood-sugar level immediately after birth.

These conditions all appear after the birth of the child. Before his birth, the diabetic mother who is being properly treated for her diabetes need not worry about him so long as she and her doctor are working together.

There is a diabetic condition, however, that is known to be dangerous to children before birth. This is called prediabetes, and is a combination of changes within the body that precede the onset of diabetes proper. These changes can all be measured, once the condition is suspected, but the danger to a child before birth lies in the fact that the prediabetes shows no outward signs that might alert a prospective mother to its presence.

After a study of prediabetic women lasting over five years, Dr. Elsie Carrington and her associates listed the situations that might reasonably lead a prospective mother and her physician to suspect prediabetes, and to test her for it. These are: (1) a family history of diabetes, (2) previous stillborn children or children who died immediately after birth, (3) oversized infants, (4) the presence of excessive amounts of sugar in the urine during pregnancy, (5) hydramnios, or an excessive amount of fluid in the amniotic sac, (6) repeated abortions or miscarriages, and (7) obesity.

It is possible to determine whether or not a woman is a prediabetic by means of a test called the glucose tolerance test. This consists in measuring the amount of sugar circulating in the blood before and after a given amount of glucose has been taken into the body.

This test should be given to every woman who has any reason to suspect that she may be prediabetic. If her condition is discovered early in pregnancy, she can be treated appropriately, and her child will in all probability be born perfectly healthy and normal. If the prediabetes goes undetected, the condition can be fatal to the child. Drs. Carrington, Reardon, and Shuman showed the importance of treatment early in pregnancy in a study of over 100 prediabetic pregnant women. Nearly half of these women were treated for their condition early in pregnancy: there were no losses among the infants in this group. The others were not discovered until much later: in this group, more than a fourth of the children died.

In another group of prediabetic women whose condition was not suspected at all during pregnancy, over a third of the infants were lost before, during, or after birth. Dr. Lars Hagbard, who conducted this study, in which he investigated over a hundred prediabetic women who had 274 pregnancies, also reported that more than a quarter of the babies born to these women weighed over ten pounds. This is in startling contrast to the 1 to 2 percent of ten-pound children born to nondiabetic women. More than a quarter of the total number of children were born before the end of the thirty-seventh week, and some of them were defective.

In view of these studies, any prospective mother who has any reason to suspect the presence of prediabetes should be tested for it immediately, in order to protect her child.

Until recently, the infants of diabetic and prediabetic women were frequently delivered by Cesarean section. There was good reason for this if the diabetes in the mother was untreated: the child was sometimes not able to withstand the stresses of birth, and the mortality rate of these children was higher than that of the general population. Now, however, when most diabetics are being treated, their children stand a much better chance of being delivered safely and normally at term. Dr. Carrington and her colleagues have proved this, and they go on to say, "The striking difference between treated and untreated patients indicates that early diagnosis is important." They conclude their studies with the comment that proper prenatal care and close observation of the child after birth are necessary, and when these recommendations are followed, the child's chances are excellent.

Serious malfunctioning of the thyroid gland in a pregnant woman is another condition that *can* affect a child but *need not* do so. When the thyroid gland secretes excessive amounts of its hormone, thyroxin, the condition is known as *hyper*thyroidism. When it secretes too little thyroxin, the condition is called *hypo*thyroidism. Many pregnant women have a mild degree of hyperthyroidism, or excessive secretion, with no apparent ill effects on their children, but in one woman out of approximately 500 the thyroid gland is sufficiently overactive so that the child may be in danger. Mildly insufficient thyroid secretion is also relatively common and also appears not to harm a child, although a severe deficiency may do so. Both conditions—hyper- and hypothyroidism—can be successfully treated, and without question should be treated.

Hyperthyroidism can be treated either by drugs alone or by drugs combined with operations for removal of part of the thyroid gland. Dr. I. H. Holt has reported cases involving pregnant women, where the treatment consisted of both drugs and surgery, in which neither the mothers nor the children suffered any ill effects. He contrasted this with the case of one woman who was five and a half months pregnant and suffering from severe hyperthyroidism. She refused to allow surgery, although her physician earnestly recommended it. He then recommended an antithyroid preparation, but she took it irregularly and stopped taking it altogether during the last month of her pregnancy. Her child, when he was born, was severely malformed.

Another hyperthyroid mother whose history was recently reported gave birth over a period of several years to three children, all of whom were also hyperthyroid. No one who has seen this family doubts that the mother's hyperthyroidism was the principal cause of the children's condition. In these cases overactivity of the mother's thyroid induces overactivity in the child's thyroid. Sometimes, however, hyperthyroidism in a pregnant woman will have the opposite effect in her child: that is, the child will be born with a thyroid gland that is not functioning at all. The mechanism here appears to be that the overactive gland of the mother takes over the function of the child's thyroid gland during the prenatal period. Thus the child's gland has too little work to do, and it fails to develop. This condition usually remedies itself within a few weeks after birth.

Most of the investigations of pregnant women with hyperthyroidism have been concerned with small groups, and for that reason are valid as pointers, rather than as established conclusions. They all lead us in the same direction, however, and unless some future investigation overthrows them entirely—a most unlikely prospect—they can stand as guides.

Primarily, these studies show that untreated hyperthyroidism in a pregnant woman is a danger to her child and that treatment reduces the danger almost to zero. In a recent report by Drs. P. Battarino and A. Capodacqua, who investigated a group of nearly fifty pregnant Italian women with untreated hyperthyroidism, the mortality rate of their children was given as 30 percent. A group of thirty women in New Orleans clearly showed the value of treatment. Among the women whose conditions had been treated, the children were far healthier than among those who had not. In addition to a high mortality rate, some of the children of the untreated mothers had goiters—which are enlarged thyroid glands—and the condition of prominent eyes called exophthalmos. This is a small group, and the subgroups are smaller still, but most investigators would agree that hyperthyroidism in a pregnant woman must be treated if the child is to be protected.

*Hypo*thyroidism, or insufficient activity of the thyroid gland, is even easier to treat than hyperthyroidism. Recent developments in this field make it unnecessary for children ever to be damaged by their mother's deficient thyroid.

If the mother's hypothyroidism is serious and untreated, her child is likely to be born with a goiter. There is also a chance that he may be a victim of cretinism, which is a condition of arrested growth and development of both the body and the mind. Cretinism is not a hopeless condition, since in most cases it responds quite satisfactorily to treatment with thyroid hormones. Sometimes, too, the child of such a mother has the same exophthalmos, or protruding eyes, that mark some hyperthyroid children.

Occasionally, in a hypothyroid mother, the thyroid gland of the child is able to compensate for her lack, and all during her pregnancy her deficiency is adequately treated by her child. The consequence of this may be that although she does not suffer from the lack during her pregnancy, the hypothyroidism returns as soon as the child is born, and the child's thyroid gland may be considerably enlarged because of the extra work it has been called upon to do. Sometimes it will return to normal size within a few weeks; sometimes it develops into a goiter.

The conclusion to be drawn here is inescapable: a prospective mother and her physician should discover as soon as possible in her pregnancy whether her thyroid gland is functioning normally. If it is not, they should cooperate on treatment that will correct the condition.

Various disorders of the blood in a pregnant woman may also threaten her child if they are left untreated. Anemia and the toxemias of pregnancy are two of these which can be successfully forestalled and treated.

Anemia is a condition in which the red blood cells are reduced in number, or the hemoglobin of the red cells is reduced, or the blood itself is reduced in amount. The symptoms are fatigue, pallor, shortness of breath, and palpitations.

Iron is one of the necessary ingredients in the formation of the hemoglobin in the blood. Before he is born, a child is not naturally endowed with iron, any more than he is with all those other necessary substances essential for growth and development. In order to manufacture his red blood cells, he must draw upon the iron in his mother's body. If she is healthy, she will be able to supply her child with enough iron, not only to answer his needs before birth, but also to store a quantity in his tissues for use until he begins receiving the solid food that will contain iron. Before birth the child must build up a backlog big enough to carry him over. If his mother is anemic during pregnancy, the supply of iron may fail somewhere along the way, and he himself may develop anemia.

The treatment of anemia is simple and direct and as effective as anything in all medicine. It consists of taking iron into the system, either through the diet or in the form of capsules or by intravenous injection. An infant's anemia can be cured by giving him iron by mouth.

In view of this treatment, a woman who allows herself to remain anemic during pregnancy is guilty of taking foolish and unnecessary chances with the health and development of her child. An example of the kind of danger she may be exposing her child to was shown clearly in a nine-year study carried out at the Elizabeth Steel Hospital in Pittsburgh. Among the many hundreds of newborn children observed there, congenital malformations were found significantly more often among the children of anemic mothers than among those of nonanemic mothers. The anemia, of course, may not be the direct cause of the malformation—poor nutrition in general is a more likely basic cause—but anemia in a pregnant woman is a danger signal, and she should heed it.

Sickle-cell anemia is a particular form of this disorder, where the defective red blood cells appear under the microscope in a sicklelike form. This disorder occurs most frequently among Negroid peoples, including the American Negro, and in some populations in Italy and Greece.

The chief danger to the mother in sickle-cell anemia is that her flow of blood may be affected by it, and in severe cases minute clots may form in her bloodstream. A local anemia may then occur in any one of the organs in her body. Toxemias, or temporary states of high blood pressure presenting a disturbance of chemical balance between body salt and water, appear more often in women who have sickle-cell anemia than in women without it. The structure of the bones may be affected, and the walls of the blood vessels may thicken. Moreover, the condition is often painful, with the pain sometimes simulating appendicitis, or heart disease, or gall bladder attack, or acute kidney disease.

Children of women with sickle-cell anemia are also seriously affected. Their mortality rate is significantly higher than that of the general population, and those who survive may be carriers of the disorder. Children who are born of mothers with sickle-cell anemia sometimes have a peculiar towering shape to their heads.

Any prospective mother who suspects that she has any kind of anemia should ask her doctor specifically to test her blood for it. Normally, he will do so anyway. Untreated, anemia can be serious for both her and her child. Treated, it will vanish.

The conditions known as toxemias of pregnancy may be due to a lack of the enzyme monoamine oxidase which helps control blood pressure. Dr. John de Maria, of the University of British Columbia, found that this enzyme was significantly lower in a group of pre-eclamptic mothers than in normal pregnant women. Although the causes are not altogether understood by medical investigators, obstetricians have made long strides in recent years in anticipating and preventing these conditions.

Toxemias of pregnancy are accompanied by the appearance of albumin in the urine and a rise in the blood pressure. In severe cases, which fortunately are rare, the mother may go into convulsions or coma, and there is some evidence that this may cause brain damage to her child.

The anticipation and prevention of toxemias of pregnancy is a matter for the physician. A prospective mother alone cannot identify the signs and symptoms that might mark her as a candidate for toxemia. She can, however, put herself under the care of a competent physician and follow his instructions closely. Good antenatal care prevents the development of toxemias.

Obesity is yet another condition that threatens the welfare of the mother and the development of the child during pregnancy. It is also a condition that can be controlled, at least to some extent, and the investigations among pregnant women who are seriously overweight underline the necessity of weight control.

In a study of 200 women, each of whom weighed over 200 pounds, 75 percent developed some kind of complication during pregnancy. Nearly half of the women developed toxemia, for example, and the proportion whose kidneys or heart became weakened was significantly higher than among the general population. Drs. H. B. Matthews and M. G. der Brucke, who carried out the study, pointed out also that the deliveries themselves were more difficult than normal: a quarter of the total number were complicated by abnormal position and presentation, and the average time of labor was considerably longer. The children, too, appeared to be affected by their mothers' overweight condition: the mortality rate among them was higher than normal, and more than half of the infants themselves were overweight.

Obesity is a complicated problem, and it is not so easily treated as some other conditions, but Matthews and der Brucke's study, and others like it, give ample evidence that obesity brings in its wake other problems. A woman who suspects that her weight may cause difficulties should discuss the matter with her physician. He may suggest postponing pregnancy, for example, until after her weight is to some extent reduced. In any case, it can be a serious problem, and should be treated as one.

The history of epilepsy provides us with a cheering example of a malady that is now understood, managed, and treated so successfully that the entire outlook for epileptics has changed. Many activities are possible to them now, for instance, that were considered dangerous in the past; chief among these is pregnancy. Dr. H. Dimsdale, who has recently studied the subject of epilepsy in women, recommends that an epileptic woman should not be discouraged from pregnancy if her attacks, when she is not pregnant, can be controlled.

Pregnancy has no predictable effect on the severity of a case of epilepsy. It has been known to cause an improvement, or at least to appear to do so. Other cases have appeared to worsen, at least temporarily. Still others seem not to change at all.

There is no evidence that a child is harmed before birth by his mother's epilepsy. Investigators have not yet determined what effect, if any, an attack in a pregnant woman has on her child. This kind of study presents obvious difficulties, but at least there is no clear connection between an event of this kind and any difficulty in a child.

In this case, as in several others, a physician's advice is the most reliable and reassuring guide. A prospective mother who is epileptic should discuss the matter fully and carefully with her doctor.

An account such as this, of the various disorders of a mother's body that may affect her child, is sobering. However much we should like to take the attitude that all women can have healthy children, simply by eating properly, exercising enough, and avoiding the obvious pitfalls, we cannot in honesty do this.

The vast majority of women nourish their children easily and bear them smoothly. Some few, however, cannot, because they are the victims of bad luck, or ignorance, or perhaps even the negligence of their own parents.

These women must take special pains in order to have healthy, normal children. It is to these women that this chapter is primarily addressed. They must carry the responsibilities that are always imposed by increased knowledge. In earlier generations, their children might have been born unhealthy, or perhaps not been born at all. We can thank our scientists for the gift of life. But we must do more than offer thanks. All of us must recognize that scientists can only show us the way; they cannot travel the path for us. To make use of the discoveries of our medical investigators, the women themselves in whose bodies the children live for the most important period of their development must actively take a part.

A woman whose own physical equipment is less than satisfactory can make a definite contribution to her own children and to the entire generation that follows her own: by understanding her problem and by taking steps to meet it, she can give to her children better bodies than the one given to her.

IX—SOME INCOMPATIBILITIES

A HEALTHY WOMAN who is able to take good care of herself and her child during pregnancy will almost surely have a healthy child at the end of the pregnancy. Once in a great while, however, a healthy, careful mother will give birth to a child with jaundice or anemia, or even to a stillborn child. This may be no fault of hers. Sometimes, if she and her doctor are forewarned, they may mitigate the effects of such an event, but at no time could the mother have prevented it.

How can this happen? How is it possible that a mother whose own body is well equipped to bear children and who nourishes herself and her child carefully and thoroughly can give birth to a child who is less than healthy?

The answer to this puzzle sometimes lies in a fundamental incompatibility between mother and child.

This is an idea that comes as a shock to some people, who see in the physical closeness of mother and child before birth a significance that it does not always possess. They are indeed close. They are dependent on each other. The activities of the mother influence to a large extent the development of the child. But they are not a single person; they are two separate individuals. And sometimes they are quite different individuals, in ways that are significantly opposed.

We are accustomed to thinking of growing children—during the adolescent years in particular—as being in conflict with their parents. They develop their own wills and their own needs, and if these happen to be in opposition to their parents' modes of thinking and acting, a family that has been a unit can develop a severe crack in its foundation.

A child before birth can be just as much in opposition to his mother as a child in high school, although in a different way. Before birth, the incompatibilities are physical, rather than emotional, and the results of them are also physical.

For example: a mother, although perfectly healthy, may be *sensitized* to certain substances. This means that these substances are particularly dangerous to her body cells, that she is unusually susceptible to them. An allergy is an example of such a condition. During pregnancy it sometimes has consequences that must be reckoned with.

If the mother's body cells are sensitized to a substance, they will automatically react strongly to that substance. In fact, they will produce antibodies for the purpose of destroying it. This is a familiar reaction: it is the mechanism by which a healthy body fights disease and develops immunity to it. Normally, the process is effective and beneficial; in fact, it is the means by which human beings survive in a world swarming with dangerous substances.

Occasionally, however, a child before birth will produce quite naturally in his body one of the substances to which his mother is sensitive. This substance may then travel through the child's system and finally find its way through the placenta and into the mother's system. The result of this would be to start a reaction in the mother's body—the manufacture of antibodies to destroy the invading substance.

So long as the reaction is confined to her body, no harm is done. The antibodies perform their function, and the mother's body continues to function normally. Sometimes, however, the antibodies cross from the mother's body through the placenta into the child's system; then the situation is changed. Here, the foreign substance is not an invader, but a normal component of his body. The antibodies from the mother's system are then attacking a healthy child, and sometimes this causes difficulties in the child's development.

Some of the most common of these difficulties begin with incompatibilities between the blood groups of mother and child. The Rhesus or Rh blood types are one of these; the common A-B-O blood groups constitute another; leukocyte, or white blood cell, incompatibility is believed by some investigators to be still another. Allergies may come under this heading too.

It is the purpose of this chapter to present what is known about these incompatibilities. The reader will soon see that the field is relatively new and that some of the work is sparse and inconclusive. Enough has been done, however, so that a prospective mother and her doctor should be aware of the various possibilities and the measures that may be taken to minimize any danger to the child.

Of the several incompatibilities, the one concerned with the Rh factors is the most complex and the most interesting. The Rh factor (more than twenty Rh factors are now known) is a particular characteristic of human red blood cells that is of special importance to prospective mothers and fathers. The factor was named for the variety of monkey, the rhesus, with which much of the experimental work in the field was done. When the blood of the rhesus monkey is injected into rabbits or guinea pigs, a special serum is obtained. This serum will cause the blood cells of some individuals to "clump," or mass together in clusters. The factor in the blood that makes it clump in response to the serum is called the Rh factor. People who have this type of blood are said to be Rh-positive; those who do not have it are said to be Rh-negative. Eighty-five percent of all white persons are Rh-positive.

An Rh-negative person will often not tolerate the presence of Rh-positive blood in his circulation, though an Rh-positive person can receive Rh-negative blood. If Rh-positive blood is injected into an Rh-negative person, or as sometimes happens when a baby's blood *in utero* enters the bloodstream of its mother and sensitizes her, there is likely to be a marked reaction. In both cases the body forms Rh antibodies which react with the baby's blood cells and destroy them. This kind of reaction is called an antigen-antibody reaction. The antigen is a foreign substance in the blood that automatically causes the manufacture of another antagonistic substance. This second substance is called the antibody. In the case of disease, the bacteria or the virus is the antigen, and the germ-destroying substance is the antibody.

In the case of an Rh incompatibility, the reaction works in this way: a mother whose blood is Rh-negative may be carrying a child who has inherited from his father Rh-positive blood. The child's blood can be considered the starting point of the reaction: the Rh-positive substances in his blood, which are perfectly at home in him, are foreign to his mother. They may be, in fact, antigens in the context of his mother's blood. If they pass through the placenta into the mother's bloodstream, their very presence can call up antibodies. These fighters then break down the red blood cells that carry the Rh-positive antigens and protect the mother's system from the foreign Rh-positive substance. So far, excellent.

The difficulty arises when these Rh antibodies pass back through the placenta to the child's system. Since they were manufactured specifically to combat the Rh substances that occur in the child's red blood cells, they now find work to do, and they proceed to attack the red blood cells in the child's body. If enough of these cells are destroyed, the child may become anemic, which is another way of saying that he has not enough red blood cells.

Jaundice is another effect. This is a condition that sometimes follows the breakdown of the red cells: other products are liberated in this process, and it is these that produce a yellowish-green discoloration of the skin. This condition is sometimes called hemolytic disease of the newborn, or erythroblastosis fetalis.

Babies with Rh hemolytic disease occur much less often than we might expect from the facts of heredity: about one in twelve pregnant women combine their Rh-negative blood with an Rh-positive baby, but Rh hemolytic disease affects only one out of approximately every 200 pregnancies. Furthermore, firstborn children are only rarely affected by this situation.

The reasons for these discrepancies are not known. Some investigators believe that the antigens in the child's Rh-positive blood sometimes fail to pass through the placenta to the mother's system. And even if they do pass through, the mother's body may sometimes fail to respond to them. Also, an Rh antigen-antibody reaction does not usually develop full-blown at the first evidence of incompatibility; to become sensitized to Rh-positive antigens evidently requires some time. In any case, during her first pregnancy an Rh-negative mother will probably not respond to her child's Rh-positive antigens. Her blood may, in rare instances, manufacture antibodies in small amounts, but not enough to cause trouble to the child's system. During this period, however, she may be developing her sensitization to the Rh-positive substance. Then, if the situation should be repeated—during a second pregnancy, for example—her system may produce enough antibodies to cause a full-scale reaction.

A transfusion with Rh-positive blood may also sensitize an Rh-negative woman. For this reason, the Rh-negative woman should never be given a transfusion with Rh-positive blood. It may produce no ill effects on her body at the time, but it may build up her sensitization to Rh-positive blood so that if she were ever to have a child of that type, a full-fledged antigen-antibody reaction might be set off. Moreover, once she is sensitized, another transfusion of Rh-positive blood could cause her death.

Damage to the child's brain is one of the possible consequences of this kind of reaction. However, this can be prevented or minimized, if the possibility is suspected ahead of time and preparations are made for one or several transfusions at birth. The brain damage is caused mainly by the deep jaundice, the jaundice substance, bilirubin, being very poisonous to brain cells. In addition, the red cells combined with the antibodies may sludge, blocking the circulation in the brain and cutting off the oxygen supply to certain sensitive brain cells. The brain, as we have seen in almost every connection, is particularly susceptible to a reduction in oxygen: brain cells can be damaged quickly and irreversibly when they are deprived of oxygen. To spell this out a little more fully: The brain damage is caused by the destruction of the infant's Rh-positive red blood cells by the antibodies manufactured in the mother's body against these cells. The antibodies pass through the placenta and attack and destroy the baby's red blood cells. Usually the main damage occurs a few hours after birth. The destroyed red blood cells release a pigment called bilirubin. If the level of bilirubin rises above 20 milligrams percent of blood, irreparable damage may be done to the child's young brain cells. The bilirubin level is measured immediately after birth by taking a blood sample from the navel cord. If the bilirubin level is elevated, or if subsequent tests done every few hours in involved babies during the first few days show a rapidly rising level of bilirubin, an exchange transfusion is done. Eighty-five percent of the infant's own blood cells are replaced by donor cells, an operation that temporarily washes out the bilirubin. Should the bilirubin level still rise, another transfusion is done. As many as five transfusions have been done during the first thirty-six hours of life. As long as the bilirubin blood level is kept below the harmful level (20 milligrams percent) no brain injury to the baby results. In short, then, the baby's red blood cells, laden with antibodies from his mother's circulation, are removed from his body at the same time that an equal amount of fresh blood is injected. Rh-negative blood is used, because of the presence of antibodies in his bloodstream; if he were given Rh-positive blood, these cells would be destroyed just as his own had been. The Rh-negative blood sustains the baby until the mother's antibodies and his own damaged red blood cells disappear from his circulation. This technique is regularly used with great success.

In the future it may be possible to prevent the development or increase of an Rh-negative mother's sensitization to Rh-positive blood. This is an interesting aspect of recent experimental work in this field. In April, 1963, investigators injected small amounts of Rh-positive blood into nine Rh-negative volunteers at Sing Sing Prison. This was done over a five-month period. Four of those volunteers were also given injections of specially prepared gamma globulin from sensitized blood donors to test whether or not their Rh antibodies might prevent sensitization to the Rh antigen. The experiment showed that the gamma globulin, consisting of concentrated Rh antibodies, did indeed provide this protection.

The investigators, Drs. Vincent J. Freda, John G. Gorman, and William Pollack, believe that the main stimulus to sensitization may come with the bleeding that occurs during labor and the delivery of the first child. If this is really true, the gamma globulin Rh antibody preparation might be used at that time and might prove to be a protection against the development of this sensitization.

Another incompatibility of blood types that can exist between mother and child concerns the four principal blood groups of human beings, which are called O, A, B, and AB. These letters were arrived at as the simplest method of referring to the two different antigens in human blood cells. These antigens are known as A and B. Some individuals have only one of these, as in blood groups A and B; some have both, as in group AB; and some have neither, as in group O.

Each of these two antigens can set off a dangerous reaction in the bloodstream of an individual whose body lacks this antigen. The mechanism is the same as that of the Rh-factor reaction. When antigen A, for example, is introduced into the bloodstream that lacks A (group O or group B), it may cause an increased formation of antibodies called anti-A. These antibodies produce clumping and destruction of the invading A-containing red blood cells. In this form, the A cells are susceptible to the anti-A's, which then proceed to destroy them.

When mother and child are of the same A-B-O blood group, no such reaction will take place. They have similar antigens in their blood, and therefore similar antibodies. There will be no clumping, and the blood cells of both individuals, even if they were to flow together through a single vein, would remain evenly dispersed in the serum.

A-B-O disease occurs almost exclusively with group O mothers. Group A and B mothers apparently produce antibodies which are too weak to harm the baby. When the mother is of group O, however, and her child has inherited group B blood from his father, a conflict may result. Observers have seen that when a specific antigen is absent from the red cells of an individual, the corresponding antibody is present. That is, this mother whose blood contains no B-type antigens would be well supplied with anti-B antibodies.

If the B antigens from the child's circulation cross the placenta into the mother's bloodstream, they may cause her anti-B antibodies to strengthen. If they become strong enough, they may then cause the child's B blood cells to clump, and may then destroy them. Occasionally, then, the antibodies will cross the placenta into the child's body and proceed to try to destroy the red blood cells there.

This reaction, like the Rh-incompatibility, occurs much

less frequently than we might expect from the numbers of women and children whose blood types are unlike. It appears three times as often as the Rh-variety, but in its common form it is so slight as to be hardly recognizable. Nevertheless, when a baby is born anemic or jaundiced, and the Rh-factor cannot be the cause, the blood-type incompatibility should be suspected.

One of the consequences of A-B-O incompatibility may be an increased susceptibility to disease in children who are born to mothers of another blood type. Several studies on this suggest that fetuses of mothers with O blood and fathers with A blood may be more subject to abortion and miscarriage than other children.

The age of the mother appears to have some connection with increased sensitization to antigens. At any rate, the older the mother, the more often we find disturbances caused by blood-type incompatibilities. Why is this so? The answer has not yet been established, but two possible reasons have been proposed. The older the mother the more opportunity there is for exposure to Rh antigens. Also, there may be an increase in the strength of the antibodies in the mother's circulation as she grows older.

An allergy is another example of a sensitization to a particular substance. Investigations of allergies in children before birth are, of course, difficult to carry out, but there are numerous examples on the medical records of allergies in children that were clearly caused before birth. Specifically, these allergies seem to be the result of the overindulgence by the mother in some particular food during her pregnancy.

A seven-month-old baby developed eczema, for example, and rapidly worsened. By the time he was fourteen months old, the eczema had become so severe that he was taken to New York Hospital to be treated for eruptions over his entire face and neck, and his scalp, and both arms and legs. On his scalp the eruptions were serious enough to have destroyed much of his hair. Dr. Bret Ratner, who saw the child and his mother, reported that the history of the child revealed two items: the mother's diet during her pregnancy had consisted largely of eggs, and the child's eczema had begun when he was first given eggs in his diet. The child's condition improved rapidly with ordinary local treatment and absolute elimination of eggs from his diet. The connection was proved to Dr. Ratner and the child's parents when, several months later, he was given by mistake a dish containing eggs, and promptly developed an asthmatic attack.

Another child patient of Dr. Ratner, a boy of five, came to the hospital with a history of asthmatic attacks covering a period of about a year. The attacks seemed to be connected in some way with food. Dr. Ratner investigated the mother's prenatal diet and discovered that she had eaten at least a pound of nuts a day during her entire pregnancy. The child's attacks began at the time in his life when he first started eating nuts. The connection was clear. The child was observed for many months after this correlation had been made, and his condition improved rapidly and remained good so long as nuts were kept out of his diet.

Allergic reactions of this kind evidently run in families. This explains why some children whose mothers eat large quantities of specific foods during pregnancy never show such a reaction, while others react violently. Not all children, in other words, are sensitized by their mothers' overindulgence. But when a child whose inheritance leads him toward this kind of allergic response is also bombarded throughout his prenatal life with a large overdose of a specific protein food, he is likely to develop a susceptibility to it and respond to it when he meets it after his birth.

Asthma and hay fever in a pregnant woman apparently have no effect on the development of her child before birth, nor even, so far as we can tell, after birth. It seems reasonable to suppose that a serious attack of asthma during pregnancy might affect a child, but the fact is that, more often than not, asthma is greatly relieved during pregnancy. The blood of asthmatic pregnant women undoubtedly carries antibodies of considerable strength to pollen and dusts, according to studies by Drs. D. H. Sherman, S. F. Hampton, and R. A. Cooke. On the other hand, the blood in the umbilical cords of their babies has no such antibodies present. Investigations of guinea pigs have proved that asthma can be allergically transmitted from mother to child, but there has been no evidence at all that this happens among human beings.

These incompatibilities between mother and child are not yet understood as well as they will be. The major studies in the field are so recent that there has not yet been time to follow them up and discover the relationships between mother and child through their bloodstreams. Yet we know enough to be reminded of an important fact that prospective mothers sometimes forget: every human being is a separate human being, even during the period of his life when he is most dependent on another.

X—INFECTIOUS DISEASES

FORTUNATELY FOR PROSPECTIVE MOTHERS, protection against infectious diseases is more effective and more common now than ever before in history, and any woman who is going to have a baby should learn as much as she can about such protection as soon as she can. Although these diseases are not always serious, they can be, to her child as well as to herself, and for the sake of the child's health and development, a mother should take advantage of every new aid that medical science offers her.

This chapter will discuss some of these infections, how they are caused, what effects they have on mother and child, and how they can be avoided.

In general, infectious diseases are caused by various kinds of minute organisms: bacteria, protozoa, and viruses. Varieties of these organisms surround us every moment of our lives, sleeping and waking, well and ill. They swarm in the air we breathe and the water we drink. They multiply wherever they find suitable conditions. And they carry disease wherever they go.

Bacteria are one-celled plantlike organisms that can apparently grow almost anywhere. These are the germs that cause tuberculosis, streptococcus and staphylococcus infections, scarlet fever, typhoid fever, undulant fever, pneumonia, syphilis, gonorrhea, and appendicitis.

Protozoa are one-celled animal organisms, and some of these, too, cause diseases in the human body. Malaria and toxoplasmosis are the most serious of these.

Viruses are even smaller bodies, which some investigators believe to be alive but which others think are composed of a complex protein with an inner core of nucleic acid.

Viruses differ from bacteria in that they evidently need living tissue in order to multiply. Viruses cause disease by penetrating individual cells; once attached to a cell, the nucleic acid in each virus disrupts the cell and causes its breakdown. Smallpox is one of the major virus diseases; happily, it is now seldom seen in medically advanced countries. Others are the so-called childhood diseases— measles, mumps, German measles, and chickenpox. Poliomyelitis, viral hepatitis, influenza in its many forms, and the common cold are also caused by viruses. Some of these can be serious to a pregnant woman and her child unless she protects herself from them; others are not.

We have only to look at the list of infectious diseases to realize how much progress has been made in recent years, and is continuing to be made, in combating them. Modern sanitation is a particularly potent force in this struggle against infection; cleanliness is our most valuable ally. Every time a human being takes a bath or even washes his hands he sweeps millions of dangerous bacteria down the drain. A city sewer system is a community-wide effort to protect itself from the germs that flourish in all waste materials.

We in the United States have got so accustomed to being clean that most of us protect ourselves in this way automatically. And our physicians have trained us to other habits, too. During an epidemic, for example, most of us stay out of crowds willingly, knowing that disease spreads from one person to another in the air between them and in the handclasp. Before the discovery of the polio vaccine, whenever polio threatened a community, the public swimming pools were deserted. We do not allow our children to put into their mouths objects they have picked up from the sidewalk. When we freeze and can vegetables from our gardens we take care that they are processed long enough to kill the bacteria that might otherwise kill us. Very few people now drink milk that is unpasteurized. We even know that although we may eat our beef rare, our pork—a favorite breeding ground of a dangerous organism called *Trichinella spiralis,* the cause of *trichinosis* —must be thoroughly cooked. We sneeze and cough into handkerchiefs instead of into other people's faces. All these things we do to protect ourselves and others from the harm that bacteria can bring to us.

In addition, we now have numerous protections against specific bacteria and viruses. The most serious of the infectious diseases—smallpox, tuberculosis, typhoid, undulant fever, pneumonia, the venereal diseases, malaria, poliomyelitis, influenza, and soon, apparently, measles and German measles—can in most cases be prevented, or at least lightened, by prophylactic means, so that their effects on pregnant women and their children are considerably less serious than they might be.

There are many cases, of course, of pregnant women who have had serious infectious diseases and who have given birth to healthy, normal children. On the other hand, many of these germs are capable of penetrating the placenta and reaching the developing child, with results that vary widely. Often, naturally, whatever result there may be is gone, or at least is unnoticeable, by the time the child is born. But sometimes a severe attack of a disease will have serious consequences. It is for this reason that a pregnant woman must take every precaution available to her.

Most of the bacterial diseases are under control in the United States, and a prospective mother who is careful and whose doctor is alert need almost never encounter any of these diseases in her own body. One or two are relatively common—infections of the urinary tract, for example—but these are also the least serious. If, against all the probabilities, a pregnant woman does fall ill with one of the more serious diseases—pneumonia, for instance —she can be treated successfully with antibiotics.

Our campaign against the tuberculosis bacterium is one of the most heartening stories of modern medicine. Earlier in our history, a "consumptive" coughed his life away helplessly; now he embarks on a routine of care that will almost always save him.

The *Mycobacterium tuberculosis* most frequently enters our systems through our noses and mouths, and the lungs are therefore usually the first seat of infection. In its mildest form, tuberculosis is a common disease. In any of our large cities, almost everyone has had tuberculosis. That is, by the time they are twenty years old 70 percent of all city dwellers show evidence in their lungs that they have at one time been infected with it. By the time they are 50, nearly 100 percent show it. Thus, tuberculosis can be a silent danger to a pregnant woman. It is entirely possible that she may have this infection in her body and not know it.

Fortunately, the tuberculosis bacterium seldom crosses the placenta to the body of a child, and as a result there are very few children born with tuberculosis, even of mothers who have the disease. Some experts warn us that this is still a serious problem and point to the steady, if tiny, trickle of reports of such children, but the fact remains that it is a problem much less threatening to any individual woman or to women as a whole than many another. Nevertheless, a chest X ray before pregnancy is a good idea, for the child of a tuberculous mother is not a good postnatal risk.

If a pregnant woman does have tuberculosis unknowingly, and if she is the one in several million who does transmit it to her child, it is likely that the child will be born prematurely, because tuberculosis tends to hasten labor. A child who is born with tuberculosis is now commonly treated with streptomycin; this saves the lives of most of these children.

Scarlet fever and typhoid can both be transmitted from a pregnant woman to her child through the placenta. Many cases have been reported of children who had both of these diseases before they were born, and recovered, or who were infected before birth and developed the diseases after they were born. Of the two diseases, typhoid is the more serious for the child: in about two-thirds of the reported cases, the result was abortion. There have been no reports of defective children as a result of typhoid.

The streptococcal infections, too, can apparently be transmitted from mother to child. Studies showing the results of these infections have been infrequent and small in scope, but there is evidence to show that erysipelas, streptococcal septicemia, and streptococcal lymphangitis are among these. Rheumatic fever is another, and this is a disease that must be regarded as serious because of the effect it may have on the heart. Dr. J. W. Ballantyne reported the transmission of rheumatic fever from the mother to her baby in the womb as long ago as 1902. And in 1909 Dr. F. J. Poynton reported large coccal growths on the mitral cusps of the heart of an infant two days of age whose mother had rheumatic fever late in pregnancy. Drs. A. Plaut and G. Sharnoff have reported the case of a seven-month-old fetus with verrucose mitral endocarditis (wartlike growth on the bicuspid valve with inflammation of the lining membrane of the heart), whose mother had "influenza" three weeks before delivery. The new antibiotics are now being used, with great success, in the treatment of these infections.

Undulant fever, a disease that originates in cows, goats, and pigs, and comes to us through their milk and milk products, is fortunately very rare now in countries where pasteurized milk is the rule. At one time it was a serious disease for mother and child, but unless we return to drinking raw milk, we need not worry about undulant fever.

Pneumonia also can be transferred from mother to child. This infection is now successfully treated with antibiotics, and the danger to a child before birth from pneumonia is slight compared to his danger a few years ago. The primary precaution here for a pregnant woman is to avoid exposure. The next precaution lies in her general health, which will enable her to avoid the infection in most cases even when she is exposed to it, and in the proper and prompt treatment of it by her physician, if she does become infected.

A bacterial infection of the amniotic cavity, where the child is developing, can in most cases be cured if it is detected in time. When it is not detected, there is a possibility that the child will be stillborn or will die soon after birth. When a mother's temperature rises and the child's heart beat becomes very rapid, a bacterial infection may be the cause; the doctor will then institute the proper treatment.

A method for determining whether or not there is a bacterial infection present in the body of a newborn child has been devised by Drs. Kurt Benirschke and S. H. Clifford. Routinely, they make a microscopic examination of a frozen section of the umbilical cord immediately after birth. They have found that if the wall of the umbilical cord is inflamed, a bacterial infection in the child is probably the cause. Sometimes the bacteria themselves can actually be seen in the small folds on the surface of the cord. According to these investigators, infections caused by viruses are not accompanied by an inflammation of the cord.

When the membranes break prematurely, the cause may sometimes be found in a bacterial infection in the child. The colon bacillus appears to be the chief offender, but staphylococcus and streptococcus infections, in particular, are also frequently associated with births during which the membranes have ruptured too soon. For this reason, a pregnant woman should always report to her doctor if the membranes rupture prematurely, so that she and her child may be treated with the appropriate antibiotics.

Infections of the urinary tract are quite common during pregnancy. Prevention of these infections lies, to some extent at least, with the individual woman. Urine is normally sterile, but when bacteria enter it, they multiply rapidly. A major step toward prevention, then, is to keep bacteria away from the urinary tract. This can be done to some degree by exercising care in toilet habits. Bacteria are always present in the intestinal tract and are excreted in great numbers with the feces. If this area of the body is not kept clean, or if it is cleaned carelessly, bacteria may be swept into the urinary tract, where they will multiply and set up an infection. The anus should be cleaned with motions away from, not toward, the vulva. General cleanliness, baths, showers, and all reasonable sanitary precautions are important. Eight glasses or more of fluid intake are also recommended.

These simple precautions are important; an infection of the urinary tract may not in itself be serious, but it may lead to a disease of the kidneys, which can be. Furthermore, the children of mothers with such infections are premature or stillborn two to three times as often as the children of mothers without these infections. This does not mean necessarily that the infections cause the prematurity and stillbirths, but it does mean that there is some connection between them.

Bacterial infections of the urinary tract may also produce iron and vitamin deficiencies in the mother, and these, in turn, may cause anemia, which can have serious consequences. The studies that brought out this fact were conducted by Drs. C. Giles and J. A. H. Brown, but these investigators do not as yet offer an explanation of the mechanism whereby the deficiencies arise.

A clear recommendation follows from all this work: young women, pregnant or not, should be regularly tested for infections of the urinary tract, and during pregnancy these examinations are essential.

Listeriosis is another bacterial infection, which is contracted either by direct contact with animals who are infected with the bacterium *Listeria monocytogenes,* or from contaminated meat or milk or water. This is not a particularly serious disease in an adult woman, even a pregnant woman, but it can be serious in a child before birth, and it should be treated promptly. In fact, it is sometimes fatal if it occurs either during the prenatal period or the first month after birth. When the infection strikes a pregnant woman as early as the fifth month of pregnancy, the child is almost always premature and stillborn. The symptoms in a pregnant woman are chills and fever and a feeling of general malaise.

In a study of thirty-four women who had histories of repeated abortions, Dr. F. Rappaport and his colleagues isolated the bacterium *Listeria* from the genital organs of twenty-five of them. The same organisms were found in the systems of their children. There seems to be no question that this bacterium was responsible for the abortions.

Women who have had several abortions or miscarriages or who have had unexplained bouts of fever should be tested for the presence of *Listeria,* and treatment should begin at once, even before the results are known. Penicillin and methoxypyridazine seem to be the most successful drugs in the treatment of this infection.

Prevention of listeriosis is partly a community problem, in the general field of public health, but any prospective mother who suspects that her meat or milk or water is in any way contaminated should report her suspicions and, for her child's benefit, avoid any further contact with it.

Syphilis is a disease that is now almost unknown among newborn children in the United States, although it is easily transmitted from mother to child. The reason for this development is the widespread use of penicillin as treatment for the disease in the mothers. If treatment is begun before the fifth month of pregnancy, fewer than 2 percent of the children of syphilitic mothers will be born with it. When the disease is left untreated, 40 percent of the children will have it when they are born. Among newborn children, syphilis can cause severe damage to every organ in the body, as well as death, and often the tragedy of untreated syphilis lies not so much in the fact that the child may die as in the certainty that if he lives, he will be seriously damaged for life.

Gonorrhea is another venereal disease than can, apparently, be transmitted through the placenta. The bacterium that causes this disease will sometimes attack a child before birth and produce gonococcal arthritis in it.

An infection caused by a one-celled animal organism, a protozoon, is called toxoplasmosis. This is serious enough to a child before birth so that all pregnant women should be tested routinely for this infection at about the end of the third month of pregnancy, and those whose reactions are negative—about 50 percent—should be tested again at about the end of the eighth month. The treatment for toxoplasmosis at present is sulfadiazine and perimethamine.

The primary difficulty with this infection is that it can exist in the mother's body without her knowing it: hence the importance of the routine tests. The only symptoms are a short period of fever, usually lasting not more than a day, and a swelling of the lymph nodes, ordinarily beginning in the neck.

At the same time, the infection can have serious results in a child. The eyes, the ears, and the brain of children before birth are all known to have been affected by untreated toxoplasmosis. The disease can be transmitted only when the organisms are actually in the mother's bloodstream. This period lasts for no more than two weeks, at the beginning of the infection; after that it is suppressed by the antibodies in the mother's system. Once a woman has been infected with toxoplasmosis, the high antibody level in her bloodstream will give her and all her later children immunity to the infection.

Malaria, although a serious disease and common in many parts of the world, is very seldom encountered in the United States, and almost never among children before birth. The malaria protozoon is capable of crossing the placenta from mother to child, and in areas where the disease is uncontrolled there are many children born with it. On the other hand, many investigators have reported cases where mothers who are immune and whose bodies are swarming with the protozoa have children whose bodies have almost none.

Thanks largely to the control of the malaria-carrying mosquito, malaria is rare in this country, but if we measure the effect of the disease around the world in the lowered vitality of large numbers of the earth's people, and its toll in the lives of children before and after birth, we must consider this one of the world's serious diseases.

Fungus infections can be transferred from mother to child during the process of birth. If the vagina of the mother is infected with fungi, and if the membranes rupture early, and if labor is prolonged, the child is exposed long enough to these agents to constitute a danger to him. In general, however, these infections disappear from the child's system within a few days.

Virus diseases differ from bacterial diseases in their effect on children before birth. Whereas a severe bacterial disease may cause abortion or miscarriage or stillbirth, some virus diseases are capable of causing developmental defects in the child. For this reason, pregnant women are particularly cautioned to avoid, as far as possible, exposure to virus diseases.

Fortunately, by the time they reach childbearing age, most women are immune to some of the more common of the virus diseases—measles, mumps, and chickenpox. An attack of German measles, however, does not give the same immunity that the other diseases confer, and for that reason the possibility of contracting this disease is particularly to be guarded against, especially during the first three months of pregnancy. The seriousness of this disease has spurred work in this field to such a degree that in mid-1962 the German measles virus was isolated and identified; the next step will be to prepare the material and technique for immunization. Other virus diseases for which various kinds of immunizations have been developed are smallpox, poliomyelitis, and influenza. Prospective mothers should discuss these immunizations with their physicians and take advantage of them. Protection of a child before birth from virus infections is as important a step as any other in his prenatal life.

We have known for generations that some of the virus diseases can be transmitted from mother to child. As long ago as 1668, in fact, the French physician François Mauriceau wrote that smallpox in a pregnant woman could disfigure her child and even cause the child's death. He knew this because he had himself seen smallpox scars on newborn children whose mothers had had smallpox during pregnancy. Afterward, particularly during the long period before vaccination became common, when small-pox was one of the worst scourges of mankind, many medical men were able to confirm this observation.

No one thought, however, to use this knowledge to explain other effects on children before birth. It was not until 1941, nearly three hundred years later, that Dr. N. M. Gregg, in Australia, put together two facts—German measles in pregnant women and cataracts in their children—and arrived at conclusions of the greatest importance. Since then, investigators have studied other virus diseases in pregnant women and have decided that many of them are occasionally guilty of harming children before birth.

This three-hundred-year lag is a remarkably long one in the history of medicine. The reason for it, however, is understandable: the most easily diagnosed virus diseases—that is, the childhood diseases—are rare among pregnant women, because most women have had them in childhood and are immune to them. Therefore there are very few cases involving children before birth whose mothers had one of these diseases—except for German measles, which does not always confer immunity.

If a pregnant woman has German measles during the first four months of pregnancy, the chances of her child being affected by it are large: one in three or four. On the other hand, if she has the same disease during the last three months of pregnancy, her child will probably be entirely unharmed by the virus.

The reason for this is that the German measles virus, like all viruses, grows particularly well on any young, developing living cells. The child, during these first few months, is growing and developing at a pace he will never equal again in his life; he is offering a perfect breeding ground to any virus that comes along. Numerous studies in the United States, England, France, Sweden, and other European countries, involving many hundreds of children, prove this beyond a doubt. In addition to the damage to the eyes noted by Dr. Gregg in Australia, the German measles virus has also caused harm to the development of the ears and hearing, and of the heart.

This disease is so serious in children before birth, and the likelihood of their being affected by it is so high, that many physicians who have studied it recommend that an abortion be performed whenever a pregnant woman has had German measles during the first four months of pregnancy. Dr. Julia Bell, of London, who has conducted many of the studies in this field, is one of these; Professor Emeritus Nicolson J. Eastman, of the Johns Hopkins University Medical School, and one of the country's most distinguished obstetricians, is another.

An additional reason for an abortion after a case of German measles lies in the anxiety that any woman would feel throughout the remainder of her pregnancy. Dr. Murray Bass has said, "Many women cannot stand the mental strain of waiting 6 months with the continued apprehension of having a defective child. Such women, in my opinion, suffer definite mental harm from this anxiety, and in these cases, with the consent of psychiatrists and a consulting obstetrician, an abortion is justified."

Measles, mumps, and chickenpox are all rare in grown women. For that reason, there are few reports on their effect on children before birth. For example, in a study of 11,000 Australians who had measles during an epidemic in South Australia, there were only 128 women of child-bearing age in the group, and of these only eighteen pregnant at the time they had measles. Eleven of these women reported that their attacks had been moderate or mild, and all their children were born healthy and normal. Among the seven who reported severe attacks, there were several children who may have been affected in various ways by it. The sample is obviously too small to allow for any generalizations. In any case, inoculations with gamma globulin after exposure to measles will lessen the severity of the attack: this is recommended to pregnant women who have not had measles and who have been exposed.

As in the case of German measles, the effects of mumps on a developing child appear to be most serious when the mother has the disease during the first three months of pregnancy. In fact, there is some doubt that the virus will affect the child at all during the later months. The studies of pregnant women with mumps are small, relatively speaking, because so few adult women ever have mumps at all; one investigator observes that pregnancy even appears to increase a woman's resistance to the mumps virus.

Reports in this area have come from Australia, Finland, Sweden, the United States, and a group of Eskimos on St. Lawrence Island. These studies add up to the general conclusion that stillbirths, premature children, and children whose development has been interfered with in various ways are more frequent among groups of women who have had mumps during the first three months of pregnancy than among the general population.

Evidence concerning the effects of chickenpox on children before birth is meager, but the little there is suggests that both child and mother can suffer from this disease. In fact, chickenpox in an adult is perhaps more serious than in a child, and a prospective mother who has not had chickenpox as a child should take special precautions against exposure to it. The largest single study of this subject was undertaken by Drs. M. M. Manson, W. P. D. Logan, and R. M. Loy, who investigated 298 pregnancies complicated by chickenpox. The proportion of stillbirths and unhealthy children was very little larger in this group than in a similar group of uncomplicated pregnancies. Nevertheless, the disease can be severe in an adult, and it can probably be transmitted through the placenta to a child, so it must be taken seriously.

Vaccination against smallpox is a routine matter in the United States now, with the result that this most virulent disease is almost completely under control here. Nevertheless, it is such a serious disease, and it spreads so rapidly among unvaccinated people, that we must remind ourselves periodically that it still lurks outside our protective shield, ready to invade with the slightest relaxation in our vigilance.

When smallpox is transmitted from mother to child it

may cause more than pockmarks on his skin; it may also cause his death. Children who have smallpox before birth are immune to the disease thereafter, and in a country ridden with it this may be some consolation. The danger of losing the child, however, is about 50 percent.

Vaccination against smallpox during pregnancy can sometimes be harmful to the developing child. It is true that during 1947 many pregnant women in New York City were vaccinated and investigators who later studied the children of these women concluded that they were probably not affected by the vaccine. However, many other investigators have reported newborn children of women who were vaccinated during pregnancy, who had clearly been severely affected by the vaccine. The cowpox virus is used for vaccinating against smallpox, and unmistakable evidences of infection with the virus have been found in the bodies of children whose mothers were vaccinated during pregnancy. Other children have been born with the typical marks on their skin; still others have been aborted or prematurely born or stillborn.

The fact that a child may occasionally be seriously damaged by vaccination of the mother during pregnancy leaves a prospective mother with but one sensible course of action. If she has doubts about her own vaccination, she probably should be vaccinated again, but considerably *before* pregnancy. In any case, she should discuss this matter with her doctor and be guided by his decision.

Poliomyelitis is a serious threat to pregnant women who have not been immunized. In the first place, these women are more susceptible to polio than women who are not pregnant. The reason for this extraordinary susceptibility has not yet been uncovered, but it is unmistakable, and when there is a question of who in a community should first be immunized, priority should be given to pregnant women.

Secondly, the children of these women are in danger, too. Figures on the exact likelihood of a child having polio when his mother contracts it are not particularly reliable; studies are scarce, and numbers of mothers and children are small, and what studies there are do not exactly agree with each other—but there is no doubt that children before birth suffer severely from the disease. The proportion of affected children ranges from 21 percent to

50 percent of all children whose mothers had polio during pregnancy.

Polio vaccine is the obvious answer here for all prospective mothers. Any physician can and will give it. In addition to helping the mothers, this vaccine will also give to the child an immunity that will last him for several months after birth.

This is an important item in any prenatal program; no prospective mother should neglect it.

Viral hepatitis is still another disease that can be serious in a pregnant woman and in her child. It produces an inflammation of the mother's liver, which is already under a strain because of the pregnancy; the combination may cause real difficulty. The virus can be transmitted to a child before birth, and there it can threaten his liver.

Vaccination against influenza is another sensible precaution for a pregnant woman. The influenza virus is not so likely to harm her child as is, for example, the measles virus, but she is much more likely to get flu than measles. At some period in her pregnancy, it is possible that influenza virus may be all around her; if this happens to be early in the pregnancy, the virus may interfere with the development of her child.

Influenza presents another problem, too. A pregnant woman may be so lightly infected with it that she is unaware of it, or even if it is a moderately severe case, she may not mention it to her physician because she "knows what to do about it." This is a mistake. Even a light case can affect her child, if it happens to come during the early period in his development when his organs are forming and growth is particularly rapid.

A far-reaching and important study of the effect of influenza on children before birth was undertaken in Dublin in 1957. Asian flu was sweeping around the world during that year, and doctors and health departments everywhere followed its progress. Dr. Virginia P. Coffey and Dr. W. J. E. Jessop, of Dublin, Ireland, predicted that the full force of the epidemic would reach Ireland in October and November. This foreknowledge gave them an opportunity to plan their study in advance.

This was an unusual and a particularly fortunate circumstance. Ordinarily, when investigators embark on a study of the effects of a virus on children before birth, they must wait until the children are born before they can even begin. Dr. Gregg, in Australia, began his study with the children who were born with cataracts and worked back through the histories of their mothers. In the case of influenza, this approach is more difficult than with other diseases, because "influenza" means different things to different people. "The flu" is an elastic term that is often stretched to cover a variety of ailments, and a woman might in all honesty report that she had had a case of influenza during pregnancy when in fact she had had a bad cold. Conversely, she could have had it and not noticed it. Both of these cases, and many in between, would cloud the results of a study of such women and their children.

In the Dublin study, these pitfalls were avoided. All the women who attended the prenatal clinics of Dublin's three maternity hospitals were examined. Over 600 women had Asian influenza at some time during pregnancy, and there was no room for any doubt about any of them; the influenza virus A was actually isolated in all of them.

These women were then matched for age, number of previous children, and A-B-O and Rh blood groups with pregnant women who had not had influenza, and the two groups of children were compared. There was no significant difference between the two groups in the number of stillborn children or premature children, but there was a difference with respect to their development. There were more than double the number of children with developmental difficulties in the influenza group than in the other, with most of the difficulties being limited to the nervous system.

Furthermore, the proportion of such difficulties was highest among children whose mothers had had influenza early in pregnancy, and it declined steadily as pregnancy advanced.

In considering these women in Dublin, a prospective mother in the United States must keep in mind an important fact: the Irish women were all poor and were almost certainly undernourished by the standards that Americans use. Dr. R. N. Acheson points this out and suggests that the damage to the children may not have been done so much by the influenza virus itself, as by the fact that these women were already taxed to the limit of their bodies to support their pregnancies. Therefore, the presence of one more enemy of whatever kind was enough to harm the children. This same situation would not prevail in an average American household where the standard of living is higher to begin with, and where most women have considerable margin to absorb such an infection. Nevertheless, similar findings in Switzerland and the United States during the same epidemic seem to bear out the Dublin study.

Two major precautions will help prospective mothers to protect their children from this virus: avoid exposure—and this means staying at home during the times when influenza is most prevalent in the community—and have the immunization against it, particularly when epidemics are expected.

These two precautions apply with equal force to all virus and bacterial infections. To avoid exposure is possibly the most sensible advice a physician can give his patient. Only by avoiding exposure can she be sure not to take any chances whatever with her child's health and development.

Unfortunately, this is not always possible. No one, least of all a pregnant woman, lives altogether removed from contact with the world. She must go to market, see her friends, and greet her older children when they come home from school. All these activities will expose her to bacteria and viruses. It is then that her vaccinations and immunizations will help her most. If she has checked these with her physician, and had those she needs, she will be safe. And, more important, her child will be safe.

Another important safeguard is general good health. If she is eating properly, sleeping enough, and not overworking, she will stand a good chance of throwing off a threatened attack of any disease. If her general health is poor, the infection may take hold and drag her condition down even further. This in turn will make her more susceptible to the next virus or bacterium that comes her way. And this will drag her down still more.

If a spiral of illness of this kind should hit her, she must, if she is eager to protect her child, go to her doctor. In fact, she should go to him with any illness of any kind, even a one-day fever that is gone before she can see him. *She should report it.* Many women feel foolish making a fuss about "nothing." But where a developing child is concerned, everything that happens is of importance, and a prospective mother should not allow her own self-consciousness to stand in the way of protecting her child. Her physician will know whether there is anything to be done; if she tells him nothing, he will never be able to help.

The medical profession has already performed miracles in protecting human beings from infectious diseases. If a mother wants to protect her child from them, she must take advantage of all the kinds of protection available to her.

XI—EMOTIONS AND FATIGUE

CAN A CHILD, before he is born, be influenced in any way by his mother's emotions?

Most women know that if a pregnant woman is frightened by the sudden appearance of a snake across the path in the woods, her child will probably *not* be born with a snakelike mark on his skin or with a reptilian character. But what about more subtle influences? If she reads poetry during her pregnancy, will her child be more likely to appreciate beautiful thoughts than if she spent her time reading murder mysteries? And in general, is it better for her and for her child to expose herself to all manner of situations, regardless of their emotional effect on her, or to sink down into a placid, uneventful life during her pregnancy?

Prospective mothers have always been interested in the answers to these questions, and some of the oldest and most persistent of the old wives' tales have to do with this subject. In a simpler day, it was commonly believed that the emotional connection between the mother and her child before birth was direct and powerful. Even when scientists came to believe—erroneously, it turned out—that a pregnant woman and her child were entirely separate, divided from each other by an impenetrable barrier, many women continued to believe the old wives' tales anyway.

We are now beginning to understand that the old wives were not altogether wrong: there is solid, scientific evidence that a strong emotion in a pregnant woman *can* produce a reaction in her child. Furthermore, scientists are reasonably sure that in some cases a prolonged emotional disturbance in a mother during pregnancy can, in some cases, cause a whole behavior pattern in a child *after birth*. It is a familiar pattern to many mothers: excessive crying, irritability, difficulties in digesting food, vomiting, and sometimes diarrhea. This pattern of behavior can, of course, also be caused by many other factors. The important thing here is the discovery that it is a pattern of behavior in the child that can be caused by emotional disturbances in the pregnant mother.

There is also some evidence, although not enough to be called proof, that a serious disturbance in a mother, particularly during the first three months of pregnancy, can affect the physical development of her child. This is the period during which the major organs are being formed, and his mother's emotional stress, like several other factors, may be able to influence the development of those organs.

A prospective mother, interested in whether or not her emotions will affect her own child, must keep clearly in mind that the discussion in this chapter concerns *some women,* whose emotional disturbances *may* affect their children. Of necessity, we are concerned here with emotions, both unpleasant and strong, and with their harmful consequences. This does not mean that all or most unpleasant emotion is harmful in its consequences. The subject is studied through these emotions because they are easier to identify and to measure than the pleasant emotions, and their results, if any, are clearer. A reader must not assume from any part of this report that her child is in danger every time she worries about something.

Almost all women who have ever been pregnant have been under severe emotional strain at some time or other during pregnancy. Indeed, it is now a well-established fact that a large number of women actively dislike being pregnant and are upset about it during a part of their pregnancy. One group of observers, headed by Dr. W. T. Tomkins and Miss D. G. Wiehl, even go so far as to say that "in every patient, no matter how 'normal,' there is some degree of unconscious conflict concerning pregnancy." They suggest that many cases of nausea and vomiting in early pregnancy can be traced back to this unconscious conflict.

In addition to these unconscious conflicts—which may

or may not be as common as the investigators suggest—most pregnant women have genuine conscious problems connected with their pregnancy, which are quite capable of affecting their emotional stability. Pregnancy brings with it changes in all a woman's relationships with other people—her husband, her parents if it is a first child, and her other children if it is not. And money, which presents a thorny enough problem for most people at all times, can become an acute worry when another child must be provided for. Even the details of living—how best to accomplish the chores, how to arrange the pleasures—frequently become more difficult. And if a young woman is not prepared for this fact (and few are) her awakening can be rude, and she can receive a severe emotional jolt.

Nor does life lessen its impact on a woman simply because she is pregnant. Accidents can happen to people she loves, illness can strike, business reverses can occur during her pregnancy just as frequently as at other times. All these events can bring emotional distress in their train.

How does it happen, then, that so many children are born healthy and normal when so many factors seem to be arrayed against their proper development?

The answer lies in the ability of almost all women to absorb distressing situations into their lives without harm either to their emotions or to their children's development. "Threshold of stress tolerance" is the scientific phrase for this phenomenon. It refers to something that every observant adult has noticed: the human spirit can absorb an enormous amount of stress without breaking. Pregnant women, in particular, whose attention is already focused inward to some extent, appear to be able to handle situations in their lives that outsiders might consider dangerous. And their children, too, are endowed with enormous resistance.

One of the primary difficulties in studying the effects of emotions is that we are still uncertain about the emotions themselves. We do not know precisely how they originate, or where. We do know, however, that in ordinary circumstances an emotion experienced by a human being cannot pass in that form from one person to another. Fear of a fire that is raging unchecked, for example, or anxiety for a husband, or anger at a friend, or grief in the face of death

—these are subjective emotions, experienced by one person alone, ordinarily unshared and unsharable.

A pregnant woman is an exception to this rule. Unlike other individuals, she can share part of her emotion with her child. This is because a portion of any emotion is physical, expressed in changes in her body that can be transferred to her child. The beginning of an emotion is the original perception—that a fire is raging, for instance, and that it constitutes a danger. The second stage is a set of physical and chemical changes in the body that immediately follow the perception of danger. We all know this sensation of change in the body that comes when we are afraid: a flood, coursing through the veins, giving the body the force to deal with the emergency in one way or another, either by fighting it or by escaping from it.

These changes are quite specific and are of two kinds: chemical substances manufactured by the nerve ends and hormones released by the endocrine glands. It is these substances—chemicals and hormones—that form the connection between the perception of danger by the mother and its effect on the growth and development of her child. When these substances enter her bloodstream, they are on the road to the child's bloodstream; when they have crossed the placenta and begin the journey through his system, he is receiving in a visible, measurable form the force that began as a subjective feeling in his mother. And it is by these chemicals and hormones that he can be affected.

An emotion, then, can be defined, for our purposes, as a strong feeling or mental state that gives rise to physical or chemical changes large enough to be measured.

Emotions come in all sizes and intensities, and a prospective mother concerned about affecting her child must remember this and distinguish among them. A continued, prolonged state of anxiety is quite a different experience from a brief period of the same feeling. Also, she must recognize that some prolonged emotional states can be deeply disturbing, while others are comparatively superficial. However, an emotion need not be prolonged to be profoundly disturbing: panic terror, for instance, is nearly always short-lived, but it makes up in intensity what it lacks in duration. The time may come, indeed, when all these various states of mind and feeling now lumped together under the word "emotion" will merit being called by separate names; they are distinct enough for that.

Any emotion begins in the brain, where a feeling is first perceived and understood. This perception causes a series of changes: first there are neurochemical changes that set off impulses, and these then pass either directly into the autonomic nervous system, or to the thalamus and the hypothalamus, where they are translated into other impulses leading to muscles and endocrine glands. All these technical words and phrases need explanation.

The autonomic nervous system is that part of the whole nervous system which leads to the involuntary muscles and glands. This is in contrast to the central nervous system, leading to the muscles of the skeletal system and under conscious control. The autonomic system, as the word implies, is independent of conscious control; it leads to the muscles of the stomach, the intestines, the heart and blood vessels, the bladder, the uterus, and all the other muscles of the body, and it leads also to the glands, mucous membranes, and numerous other organs that work without direction from the conscious mind.

This is a powerful and extensive system. When it is functioning normally, it regulates the activities of all these organs and systems, and the body as a whole is pervaded with a feeling of well-being. When, for example, conditions of stress and danger prevail, the nervous system is activated and the nerve endings immediately liberate one or more chemical substances. One of these substances is sympathin, which closely resembles adrenalin. This may enter directly into the bloodstream, travel by this route to the placenta, and finally find its way into the bloodstream of the child.

Impulses from the brain also race to the hypothalamus. This organ is situated at the base of the brain and is, in effect, the coordinating center of the autonomic nervous system. It regulates the entire system of ductless glands and through them has a powerful effect on the entire body.

For example, the pituitary gland secretes, among numerous other substances, the adrenocorticotrophic hormone, or ACTH. When this is liberated, as a result of a signal from the hypothalamus, it stimulates other glands to action, particularly the adrenal glands. These glands secrete cortisone. And all these hormones are capable of passing through the placenta to the child.

Thus, the end product of the original state of fear in the mother may be cortisone in the circulation of the child in quantities that may be harmful.

This is a far cry from the belief that a child's skin can be marked because of something his mother saw during her pregnancy. On the other hand, it is not far from the observation, made by generations of women, that some individuals tend to have calm babies and some tend to have excitable babies. And who can say how much of these qualities is caused by inherited traits and how much by their environments? A woman who has a difficult time during her pregnancy, whose life has been peppered with worrisome details and accidents, is quite likely to have a difficult child, one who is overactive and has trouble with his food and his elimination, who cries more and sleeps less than normal. Conversely, a woman whose life is serene and placid, who seems to float through her pregnancy with no difficulty and with very little emotional upheaval, may very well have a child who is serene like his mother. And the connection may not be entirely through heredity; it may be partly, at least, through the uncontaminated environment that a peaceful woman provides for her child before he is born.

Evidence about children before birth is naturally difficult to obtain. Most of the evidence we have rests on two forms of behavior: the child's movements and the beating of his heart. These are simple, direct actions, reasonably well understood, and measurable. Furthermore, there is enough variation in them so that they can constitute an effective yardstick. When an investigator observes closely the movements and heartbeat of a child before birth, and correlates the changes in these two things with what he knows of the emotional condition of the mother, he is beginning to understand the connection between the two. Then, when he studies the same child after birth, and observes his behavior and health, and correlates these facts with what he already knows about the child's prenatal life, he begins to see the train of events as it actually happens.

The investigator can also sometimes arrive at tentative conclusions by starting with newborn infants who show certain clearly defined forms of behavior, and by tracing these back into the histories of their mothers' pregnancies. In this way anxieties and stresses that can be correlated with the children's behavior are occasionally brought to light.

The movements of children before birth have been divided, for the purpose of labeling and studying them, into three kinds: a slow squirming movement, a small rhythmic movement called hiccups, and a sharp kicking movement. These were classified in this way by Helen Newberry Norman, working at the Fels Research Institute at Yellow Springs, Ohio. The group of investigators at this laboratory, under the direction of Dr. Lester W. Sontag, has done more than any other single group to add to our knowledge of the effects of a mother's emotions on children before birth.

Kicking movements vary from mild to convulsive. Possibly the earliest account of a connection between a mother's emotional state and a child's movements was recorded in 1867 by Dr. James Whitehead. He described a woman in her ninth month of pregnancy who nursed her twenty-month-old child, an only child, through a severe three-week attack of a serious disease. As soon as it became clear that the child would live, the mother collapsed, exhausted. The child in her womb then began to kick so violently and for so prolonged a period that Dr. Whitehead called it a convulsion. The kicking grew more and more severe for several hours. Dr. Whitehead gave the mother chloroform and "nepenthe," and the kicking began to subside. From beginning to end, the attack lasted for five hours. Twenty-one days after this, the baby was born, healthy and vigorous, and for as long as Dr. Whitehead observed him—a matter of thirty-five days—he showed no tendency whatever toward unusual behavior.

Dr. Whitehead was well ahead of his time in his understanding of this episode. He pointed out in his report that although a child before birth may be quite unperturbed and unharmed by a physical shock such as its mother falling from a height, "It seems to be otherwise when the mental system of the mother becomes unbalanced by violent and sudden shocks of anguish, or by prolonged and severe anxiety."

Measurements of a group of emotionally undisturbed pregnant women have been the starting point of some contemporary investigations. Since the emotions act largely through the autonomic nervous system, Dr. Hudson Jost, of the Fels Institute, correlated the activity of a number of children before birth with the activity of the autonomic nervous systems of their mothers. These last measurements included the resting heart rate, the respiration rate, the variations in both of these, and the conductivity of the skin.

The first observation was that there is wide variation among women in the activity of their autonomic nervous systems. That is, the heart and breathing rate and the conductivity of the skin is much higher in some women than in others. And Dr. Jost's observations showed that the women who had the highest autonomic activity— that is, the fastest heartbeat and respiration rate and the highest skin conductivity—also had the most active children.

Some women show a great variation in themselves, too, in this autonomic activity. Their heartbeat is much faster at one time than at another. The study showed that those women in the high autonomic group with the greatest degree of variability in their autonomic patterns had the most active children.

Measurements were also made of these same patterns at different periods of pregnancy. In almost every case, there was a significant drop in heart rate, respiration rate, and skin conductivity during the last month of pregnancy. At the same time, there is almost always a decrease in the activity of the child. During the last month before birth, he is much quieter than he has been heretofore.

The investigators—again Dr. Lester W. Sontag and his collaborators at the Fels Institute—gave special attention to the women who were undergoing emotional stresses. They learned that during the periods when the mothers were disturbed, the body movements of the children increased by several hundred percent. Even when the disturbance lasted only a short time, the increased activity of the child continued for several hours, and when the stress endured for several weeks, the activity of the child over the entire time was greatly increased.

One result of increased activity such as this is to lower the birth weight of the children. Even when the length of the children is the same, the weights of the active children are lower than those of the placid children. Dr. Sontag pointed out that the child who is very active is simply demonstrating a common method of reducing: he exercises without taking in any additional food. Since he does not use the mineral part of his food for energy, this goes into the growth of his skeleton, and he is as tall as he would be under any circumstances; but he burns up in physical activity what would otherwise have gone into the deposit of body fat, and therefore he weighs less.

A typical infant who has been influenced by his mother's severe emotional stress is graphically described by Dr. Sontag.

Such an infant is from the beginning a hyperactive, irritable, squirming, crying child who cries for his feeding every two or three hours. . . . Because his irritability involves the control of his gastro-intestinal tract, he empties his bowels at unusually frequent intervals, spits up half his feedings and generally makes a nuisance of himself. He is to all intents and purposes a neurotic infant when he is born—the result of an unsatisfactory fetal environment. . . . In certain instances of severely disturbed maternal emotions which we have observed— for example, one in which the father became violently insane during his wife's pregnancy—the infant's bodily functions were so disturbed that a severe feeding problem resulted. The child was unable to retain food and became markedly emaciated and dehydrated.

An unmarried woman of thirty-five was another case described by Dr. Sontag. "She had suddenly learned that the father of her child, to whom she had expected to be married, was already a married man. To add to her distress, her very severe family made life entirely miserable for her." When her child was born, he was among those who were hyperirritable and unable to eat.

In eight cases studied by workers at the Fels Institute, where emotional distress was an obvious factor and had existed over a long period of time, seven infants had feeding difficulties and vomiting. Some of the children had frequent stools for several months.

A group of symptoms in infants—restlessness, excessive crying, irritability, vomiting, and frequent stools—was observed by Dr. E. K. Turner, in Australia. She noted that this group of symptoms was particularly frequent among illegitimate babies, and she believes that the mothers of these children are likely to be under considerable stress during their pregnancies, and that this stress causes these symptoms.

A study in San Francisco, reported by Dr. Antonio J. Ferreira, showed overwhelming evidence of this connection. During the last four weeks of pregnancy, nearly 200 women at Letterman Army Hospital answered a questionnaire that had been designed to discover whether they disliked being pregnant or were in any way unhappy about their expected babies. The babies of all the women were then carefully observed, and their behavior was judged in each case to be either "upset" or normal. There were twenty-eight children labeled "upset," and 135 normal. The two scores were then compared, with interesting results: there was a definite correlation between the women who had been unhappy about pregnancy and the babies who were "upset."

A pregnant woman's emotional state may also have some effect on her ability to absorb nourishment. Dr. Genevieve Stearns conducted a study to investigate this connection. She worked with a group of unmarried pregnant girls and young women between the ages of thirteen and thirty. She found that all the girls and women who were emotionally stable absorbed and retained more nourishment than was needed by their children. The situation was quite different, however, among those who were emotionally distressed; each time they were disturbed their absorption and retention of food was seriously affected. One girl, who was the most continuously disturbed by her pregnancy, lost calcium steadily during the course of the study. Her total calcium loss was at least double the amount used by the child. And her losses of phosphorus and nitrogen left her body with a serious lack.

Another connection, suspected but not yet proved, was that between a mother's emotional condition and a small number of children with cleft palates. Over 200 mothers of children with cleft palate or harelip were studied in this investigation by Drs. L. P. Strean and L. A. Peer. They questioned the women to try to discover whether their medical histories showed any single factor common to all of them that might account for the cleft palates in their children. Their conclusion was that severe emotional stress, accompanied by vomiting, was the one most common episode among the women. Twenty-five percent of the mothers also said that some member of their family had cleft palate or harelip. This led the investigators to the observation that possibly—in fact, probably—both an inherited tendency toward cleft palate or harelip *and* emotional stress at the critical period were responsible for these cases.

Some of the most interesting work in this field has been done with laboratory animals. Since it is impossible to carry out many of the important investigations with human beings, it has become necessary to study the effects of emotions on the growth and development of other mammals. The results of this work are not transferable in every detail to human beings, but human beings share many qualities with other mammals, and an understanding of our problems comes to us sooner and more completely if we are able to study them also in other mammals.

For example, the emotional qualities of young rats have been carefully studied, to determine whether or not a strong anxiety in their mothers during pregnancy was transferred to the young in the womb. The investigators concluded that there was no doubt that when the mothers were particularly anxious during pregnancy, the young themselves showed much more emotionality than other animals. Furthermore, these qualities persisted to some extent into adulthood.

Another study, carried out by Dr. F. C. Fraser and others, demonstrated that when cortisone is present in the body of a mother mouse between the ninth to the fifteenth day of gestation, a great many of the offspring of that mouse will be born with cleft palate. Drs. T. Ingalls and Cruley have demonstrated this same fact, narrowing the period to only two days—the twelfth and thirteenth days of gestation. These are clearly the critical days in the development of the palate in the mouse, and this work proves beyond any reasonable doubt that if an unusually high amount of cortisone is circulating in the mother's bloodstream during these days, the cell division that would normally take place in the bones of the head and jaw is sometimes interfered with, and some of the young are born with incompletely developed bones.

Drs. Strean and Peer, who studied the mothers of the children with cleft palate, added one of the most interesting of the studies in this field when they enlarged their work to include laboratory animals. Their subjects were a group of mice that had been treated with cortisone during the critical period of the formation of the palate and could be expected to produce a great many young mice with cleft palate. These mice were fed a diet containing a large amount of the various vitamins that counteract the effects of cortisone. In other words, this was an antidote to the cortisone. The result was a much smaller number of offspring born with cleft palate.

This may conceivably be the forerunner of studies that will eventually show us how to control the effects of emotion on a child before birth. It would be ironic, indeed, if even before we fully understand the connection between emotion in the mother and the development of her child, our medical research men will have worked out techniques to circumvent it.

Although medical researchers will be studying this subject in detail for many years to come and will continue to add to our knowledge of it, for practical purposes we already know enough to perceive the trend that these future studies will take.

This much is clear: the emotions of a pregnant woman do have an effect on her child before he is born, in the form of chemicals and hormones sent to him from her circulation through the placenta to his bloodstream. It is not usually a harmful effect; sometimes it results in nothing more than an increase in his activity, which in turn may cause a slight decrease in his weight at birth. But it *can* be harmful to his development if the emotion is severe, if it happens to come at a critical period in the development of the child, and perhaps also if it happens to coincide with some inherited tendency.

But what is a prospective mother to do about it?

The difficulties that face any pregnant woman in her efforts to protect her child from these possibly harmful effects are obvious: she has not the control over her emotions that she has over the food she eats, for example, or whether or not she smokes. Moreover, some emotional distress is inevitable in most lives, and a woman who lives responsibly, committed to the world around her and responding to it, cannot suddenly turn it all off when she becomes pregnant.

But there are several precautions she can take. She can concentrate on eating properly, on avoiding infections as much as possible, and avoiding undue fatigue. These three precautions are basic in establishing her general health and state of mind, and if she is successful in accomplishing this much, she will go a long way toward controlling her emotional condition. Nutrition and infection have already been treated in these pages; fatigue has not, and a discussion of this subject properly belongs here. A woman who is well nourished, free of infection, and—equally important—free of fatigue, is the most likely of all women to be emotionally stable. Fatigue is an archenemy of stability.

FATIGUE

"But I'm *always* tired," says the young pregnant woman who keeps house, launders and irons, and takes care of her husband and several children. And she is right: fatigue is the occupational ailment of young mothers.

The young pregnant woman who has no children, but who works at a full-time job and keeps house during the evenings and weekends, agrees. She, too, is tired. Both women will become more tired as their pregnancies advance.

This kind of fatigue is understandable and reasonable. Women with families of young children and women who work full time, in addition to keeping house, work hard and for long hours; it would be surprising if they were not tired.

A certain amount of this kind of reasonable fatigue is apparently not harmful to either mother or child—at least there is no evidence that a child before birth reacts to fatigue in the way he reacts to his mother's emotional distress: his heartbeat is no faster, nor does he kick harder or more often when his mother is tired than when she is rested. And millions of women have worked hard during pregnancy without any apparent effect on their children.

Nevertheless, it is possible for a pregnant woman to work *too* hard for the good of her child. There are more premature and stillborn children, for example, among women who work at full-time jobs away from home during pregnancy than there are among women who stay at home. And there is no question that a woman who is exhausted is an easy target for emotional disturbances.

Most of the scientific studies of fatigue in pregnant women compare women who stay at home with women who work at jobs away from home. These studies showed that women who worked the longest period had the least healthy babies.

It seems obvious that the kind of work these women do, either at their jobs or at home, would be an important factor in their fatigue and in the rate of prematurity of their children. Dr. A. D. McDonald studied over 3,000 women in England in 1958 and came to the conclusion that women who did the *heaviest* work had the highest percentage of affected children. Dr. McDonald particularly noted laundry workers, 15 percent of whom gave birth to children with various defects. The children of women who did light work, even away from home, were much closer to the figures for the general population.

Muscular activity, then, may be a key to the difference between the children of working and nonworking mothers. Dr. N. Morris and his co-workers point out that muscular activity tends to reduce the circulation in the uterus of a pregnant woman, and rest tends to increase it. As we know, anything that interferes with circulation, either in the mother, in the placenta, or in the child, can produce anoxia in the child and affect his growth and development. Fatigue, then, may not be the only agent at work here; impaired circulation may be another.

On the other hand, we cannot even be absolutely sure of that; some muscular activity by prospective mothers appears to be beneficial to children before birth. Dr. G. L. Erdelyi reported in 1960 on a study of 172 women who were among the best athletes in Hungary. They were compared with an equal number of nonathletic women. The athletes had fewer complications of pregnancy, shorter labor time, and only half the number of Cesarean sections than the other women. These results are difficult to appraise: athletes are likely to be in magnificent health, they are accustomed to considerable exercise as a part of their daily lives, and as a group they know how to take care of their bodies. Their superior performance as mothers may not be due to exercise alone, but may be a reflection of their general physical excellence, as well as other correlated traits.

At the other extreme, mothers who spend some time in rest homes before their children are born have larger and healthier children than mothers of similar social and economic background who do not. Beginning in 1918 Dr. Sigismund Peller, who has studied thousands of pregnant women in and out of rest homes in both Vienna and New York, has concluded that the children of women in rest homes are stillborn, premature, or ill less frequently than children of women who live at home. He points out that to live in a rest home means better food for many women than they get at home, and for all of them it means detachment from their daily worries, emotional stresses, and hard physical work. During the period 1939 to 1948, when Dr. Peller studied these women in New York, the mortality rate of the children of all white women in the city was 4.4 percent. The mortality rate for the women in rest homes was only 2.4 percent.

Poor habits of posture during pregnancy may result in stress and excessive fatigue. Uncomfortable shoes, for example, are a common cause of fatigue. Sandals and heelless shoes place an excessive weight on the heels and tend to produce sway-back posture with resultant lumbosacral strain. The recommended shoe is one that should be modeled to the profile of the foot, with a 2 to 2½ inch broad-based heel.

The correctly shod woman when positioned against a wall, with head erect and feet parallel, should be able to stand with the back of her head, upper back and shoulders, buttocks, and back of her heels in contact with the wall. Her lower back curvature should, ideally, be a hand's thickness from the wall. Exercises of heel-raising and lowering in this position, as well as abdominal breathing by contracting the abdomen from below upwards, are recommended.

In studying the effects of both emotional stresses and fatigue on pregnant women and their children, the great lack is on the quantitative side. This is not true of nutrition, for example, where we know just how much vitamin C is needed in a normal human body. We cannot measure either emotion or fatigue. When a pregnant women gets a scare from a near miss in an automobile, is her emotion mild, moderate, or severe? There is no way of telling. And she knows she is tired, but she cannot say *how* tired she is, except in the broadest manner.

Only one general recommendation can be made about fatigue: a pregnant woman who has a job outside her home should probably stop working during the fourth month of pregnancy. Beyond this, our research workers have not been able to go.

It is all the more important, therefore, that the prospective mother keep an eye on her own inner workings, with respect to her emotions and her state of fatigue, and order her life in such a way that she is exposed to highly emotional situations as rarely as possible, and that she concentrate on avoiding fatigue as relaxedly and as assiduously as she concentrates on avoiding malnutrition.

Even her doctor cannot help her much here. He is no better judge of her emotions and her feelings than anyone else. She can talk it all over with him, and she should probably do this, if she has any doubts about her condition. But in general, this is one job a prospective mother must do for herself.

XII—SOME PROBLEMS:

PREMATURITY, POSTMATURITY,

CESAREAN SECTION,

MULTIPLE BIRTHS

A PREMATURE BABY is a baby in danger, and a prospective mother should take every possible precaution to avoid having her baby early.

Throughout this book, prematurity has appeared in almost every chapter, cited as an important result of malnutrition, oxygen starvation, some illnesses, drugs, smoking, emotional disturbances, and noninfectious disorders. In fact, any interference with the normal and natural workings of the process of gestation may cause a baby to be born prematurely.

It is now time to consider the premature children themselves. Precisely what is wrong with being born early? Many a pregnant woman has said that the last month of pregnancy is the longest month by far, that she would gladly welcome an early end to her pregnancy. And the care of premature children is now so successful that most mothers see little danger to their children in an early arrival.

These women are ill-informed. A child born at term can deal with conditions and events that would endanger the very life of a premature child. The protections against infections, for example, that all of us carry within us are not always present in a premature child. He often fails to run a fever when he is ill, or even to muster an increase in his white blood cells to combat an infection.

Sometimes he is too small and too weak to suck. He is particularly susceptible to cold, and shock, and physical handling.

If his development had proceeded normally, he would still be lying in his mother's body, warm, protected, nourished, putting all his energies into perfecting his organs, developing his enzyme system, and generally readying himself for birth. Instead, his energies must be put into the struggle to breathe, to suck, to digest his food, to eliminate the waste, and all with not-quite-finished equipment. In short, he must work hard simply to stay alive. Sometimes he can't make it: the mortality rate of premature children is twenty times that of children born at term.

"But I can't stop my child from being born early," says a young mother. "Once he starts coming, there's nothing I can do."

She is right. Once he starts coming, it is too late. But every pregnant woman can do several things during the course of her pregnancy that will greatly decrease the chances of prematurity.

The first and most important of these is to be sure that her diet is good and plentiful. Malnutrition brings on prematurity; therefore good nutrition helps to prevent it. In groups of pregnant women who have been given supplementary diets, the prematurity rate has dropped dramatically.

Next, she can stop smoking. As we saw in our discussion of the effects of smoking on the development of children before birth, there is no question that the more a mother smokes, the greater the risk of her child being born prematurely.

Also, she can avoid fatigue. This means overwork, not normal tiredness after a day's work, which disappears with a night's sleep.

In general, the best recommendation for avoiding a premature child is again the simplest: give a child the best food available, enough oxygen, and interfere as little as possible with his development.

A premature child is defined as an infant who weighs less than 5½ pounds at birth. Between 5 and 10 percent of all children born each year in the United States are premature; in round numbers this is between 200,000 and 400,000 children. Some of these children are born so early and are so small that their chances of survival are negligible. The greater the child's weight—that is to say, the closer he is to his normal birth date when he is born—the better is his chance to live.

Prematurity is not necessarily a permanent handicap, and most of the children who survive a premature birth grow to be normal children and adults. In fact, some of the world's great figures—among them Newton, Darwin, Napoleon, Cardinal Newman, Voltaire, and Victor Hugo —are said to have been born prematurely. And even persons who have been severely handicapped by their premature births frequently grow into men and women of high achievement.

On the other hand, prematurity is definitely a handicap during infancy. The basic functions of premature children operate less smoothly than those of children born at term. Most hospitals have special nurseries for premature children, where they are given the help they need in order to survive—oxygen, warmth, frequent feedings—and this care saves the lives of many of them. But they appear to be afflicted with more maladies during their first few months than are other children. For example, 50 percent or more of premature children are born with hyperbilrubinemia, or jaundice. And the frequency of this condition is directly correlated with the degree of prematurity. That is, the larger and older premature children run less risk than the smaller and younger. In a study in England of nearly 600 premature children, Drs. M. E. Barton, J. Walker, and W. Walker divided the children into two groups, those whose birth weights were between 4 pounds and 5½ pounds, and those who weighed less than 4 pounds. Jaundice appeared in 2½ times as many of the lighter children as the heavier.

Premature children are also susceptible to myopia (nearsightedness). Among a group of sixteen premature infants, Dr. N. L. McNeil found that twenty-two out of the thirty-two eyes were myopic, although not one of the children had myopia in his family history.

Prematurity and cerebral palsy also have a connection. Among children with cerebral palsy, a high percentage were born prematurely. In a group of over 500 palsied children in New York State, Drs. A. M. Lillienfeld and B. Pasamanick found that 22 percent had been born prematurely. In another similar group in England and Wales, Drs. L. Greenspan and G. G. Deaver discovered that 35 percent had been premature.

The blood vessels in the brain of a premature child are more permeable and more fragile than they are in full-term babies. This represents another danger, because a sudden change in pressure can lead to the breaking of a vessel, resulting in cerebral hemorrhage or fatal brain damage.

Premature children are susceptible, too, to a plugging of the intestinal tract with meconium. This is a waste material which in a fully developed child is discharged soon after birth. A premature child sometimes lacks the enzymes in his system necessary to break down the proteins in his stomach; in addition, his intestines are sometimes not mature enough to have begun the smooth contraction and relaxation that propels the contents along the tract. These two conditions occasionally combine to cause difficulty in a premature child.

The development of teeth may sometimes be affected by prematurity. There is some evidence that the teeth of prematures are more susceptible to caries than the teeth of normal children. In Glasgow a group of five-year-old children who had been premature were studied by Drs. J. Houston and J. Aitchison; these proved to have nearly twice as many carious teeth (10.3 teeth) as a group of London children of the same age (5.8 teeth) who had been born at term. The investigators also found malocclusion in 16.6 percent more of the premature children as compared with 10 percent in the general population. In a Swedish study, premature children were found to have defective enamel on their teeth more frequently than full-term children.

One of the most thoroughgoing studies of premature children has been carried out by Dr. Cecil Mary Drillien of Edinburgh. She found that most of the premature children in her study eventually caught up with other children in their mental and physical qualities; this was especially true of the group whose birth weight fell between 4½ and 5½ pounds. Nevertheless, among the prematures as a whole there was a disproportionate number of children who were handicapped, small, and more susceptible than normal to disease. For example, of twenty-three severely handicapped children in her study, only one weighed over 5½ pounds at birth. And the number of children hospitalized with illness in the first two years of their lives was twice as high in the premature group as in a similar group of full-term children.

In a later study, Dr. Drillien came to the conclusion that prematurity is sometimes also accompanied by mental deficiency; the proportion of children who were below average in intelligence was greater in groups of premature children than in similar groups of children who were born at term. Moreover, the level of intelligence seemed to have a direct connection with the degree of prematurity: the lower the birth weight, as a rule, the lower the intelligence.

Although premature children, in general, do catch up with other children eventually, their path is rougher, and they present their parents with more behavior problems than do other children. Dr. Mary Shirley, at the Harvard Child Study Center in Boston, described premature children as being slower to talk and to control their muscles than full-term children. They achieve bowel and bladder control later and with greater difficulty. Their attention span is shorter and they tend to be more emotional, jumpy, anxious, and shy. Dr. Shirley explains this combination of symptoms by pointing out that "premature births often are cataclysmic, unduly prolonged or precipitant, both of which conditions subject the baby to birth trauma."

After birth the premature child loses weight over a longer period than full-term children. His energies are so completely absorbed by the job of maintaining his essential functions that he has none to spare for growing.

As these children grow into childhood and adolescence, up to the age of twenty, they are commonly shorter and lighter in weight than full-term children. Some investigators believe that they are also more likely to have personality difficulties.

Nearly 1,000 children—half premature and half full-term—were studied at the age of forty weeks by Dr. Hilda Knobloch, who came to the conclusion that the full-term children had a rate of mental deficiency only one-half that of the premature children. And she, too, found that the lower the birth weight, the higher the proportion of handicapped children. Dr. Knobloch and her co-workers said:

> The data . . . indicate that there is a relationship between neurologic status, physical growth, physical defect, and illness, and suggest that factors responsible for prematurity, and which cause cerebral damage, have a generalized deleterious effect. . . . These findings lend . . . support, if such is needed any longer, to the necessity for preventing prematurity in order to effect substantial reductions in the incidence of physical and neuropsychiatric disability.

A single fact stands out clearly from all these studies: prematurity is inseparable from many other factors; it is a result of some, it is the cause of others, and it appears to accompany still others. Very young mothers, for example, and women who are having their first child after the age of forty, give birth to more premature children than women in their twenties and thirties. This seems to mean that the reproductive systems that are not yet fully developed, and those that have begun to age, are in general not so hospitable an environment as those that are at the peak of their efficiency. Women who smoke have more premature children than nonsmokers. Women who have complications of pregnancy, those with a history of miscarriage and abortion, and those who have twins or triplets also have more of their children born prematurely than other women. Hypertension in a mother sometimes accompanies a premature child. Syphilis and heart disease are also found more frequently among mothers of premature children than among mothers of full-term children. And groups of women who live in poverty, with poor nutrition, produce more premature babies than more fortunate women.

In such a complex interrelation of cause and effect, it is impossible to assign any single factor as being responsible for the premature birth of a child. A poverty-stricken mother, undernourished herself, exhausted and on the verge of illness much of the time, is quite likely to have a premature child who is also weak, susceptible to illness, and perhaps even defective in some of his organs or functions. In the case of this child, can we say that any one factor is responsible for his prematurity? Or can we say that his prematurity alone is responsible for his childhood illnesses or his slowness in school? Obviously not: many threads are woven together in any individual; when prematurity is one, it is attached at both ends to several others. We must beware of oversimplification.

An object lesson in this has been offered to the world through a revealing study conducted in England between 1956 and 1960 by Dr. J. W. B. Douglas, in which he showed that many of our previous beliefs about the causes and results of prematurity were too superficial. This study is worth describing in some detail, as a good example of imaginative research.

Dr. Douglas studied in detail over 1,300 children, all of them eight years old, half of whom had been premature and half full-term. He tested them all in reading, vocabulary, and general intelligence, and compared their scores. He was not surprised to find that the premature children scored lower, on the average, than the full-term children. He then expected to discover that those children who were smallest and lightest at birth would be the ones with the greatest handicaps; he was startled to find that this was not true. Furthermore, relatively high scores were made by many premature children whose mothers had a history of toxemias of pregnancy; relatively low scores were made by many whose mothers had had no abnormalities of pregnancy. These findings were so unexpected that Dr. Douglas and his colleagues decided to continue the investigation.

When the children reached the age of eleven, their records were compared again. This age was chosen because in Great Britain most school children take examinations at the age of eleven to determine whether or not they will go on to secondary schools. Also, more grades from school were available, more scores from intelligence tests, more reports from teachers. When all these scores were compared, the children who had been premature again scored lower.

But Dr. Douglas was not satisfied. He went on to compare other factors in the lives of these children. He chose three: social and economic background of both parents, whether the mother of the family took satisfactory care of the children, and whether the parents were interested in the child's progress in school. These, of course, are all factors in the child's environment and have nothing to do with his innate intelligence.

This brought to light some revealing facts: the premature children who came from good homes scored as well as the full-term children who came from good homes. The premature children who came from inferior homes scored lower than the full-term children who came from good homes. And the premature children who came from good homes scored *higher* than the full-term children who came from poor homes.

These findings shift the spotlight from the fact of prematurity—an oversimplified explanation—to the more complex social and economic factors that appear to be responsible for *both* prematurity and lowered intellectual achievement. Only a quarter of all the families of the premature children were rated favorably according to Dr. Douglas's three criteria; it seems clear, therefore, that most of the children who were premature were also victims of poor family background, and that we are guilty of ignoring important factors in the development of these children, if we blame their poor school records on their prematurity alone.

A review of the reported studies carried out by Dr. I. Alm, in 1953, showed similar results. Dr. Alm concluded from these studies that premature children catch up with full-term children physically and mentally by the time they are two or three years old, unless they have had injury to the brain. After that, they are no different in themselves from the children who were born at term. He suggests, as others have done before and since, that some of the differences between premature and full-term children in later life are probably due to poverty rather than to the fact of prematurity.

This suggestion points directly to a solution to the problem: better maternal care during pregnancy. This is no easy solution, and we cannot expect to erase altogether the results of poverty, neglect, ignorance, and indifference. But we can take a large step in the right direction simply by providing good diets for all expectant mothers. This has already been done on an experimental scale, and the results are definite enough to constitute proof that it will work.

In England and Wales, the number of premature children and childhood deaths has dropped significantly as a direct result of supplementary diets given to pregnant women. In Finland, the prematurity rate has been lowered to 2.4 percent, as compared with 5 percent in the other Scandinavian countries, and between 5 and 10 percent in the United States. This was done in Finland by a program for pregnant women that included a good diet, avoidance of heavy work, and frequent rest periods during the day.

Poverty is not the only cause of prematurity. Dr. C. Räihä studied some 2,000 pregnant women in Finland, paying particular attention to the volume of their hearts. He then followed them through delivery and examined them again. He also observed their children. From his results he concluded that the smaller the volume of the mother's heart, the greater was the risk of her giving birth to her child prematurely. This fact is useful to know, even though the reasons for it are not altogether understood. A woman with a heart much smaller than normal can now be identified as a special problem and receive special care. Other studies of this kind will extend this useful information.

An individual woman contemplating the complex problems that include malnutrition, illness, emotional disturbance, anoxia, prematurity, and the unfortunate children born into them, may respond by feeling overwhelmed. How can she, as an individual woman, hope to do anything about them? When the causes of all these things are deeply rooted in poverty and ignorance among millions of people, she alone can do nothing.

And yet, she can learn the facts as they become known and apply them to her own situation. She can conduct her life during pregnancy according to the information that arrives in a steady stream from our laboratories. She can take seriously the proven connection between smoking and prematurely born children. She can eat sensibly, in spite of temptations to eat foolishly. She can try her best to avoid overwork and fatigue. If she does all these things, she will be doing what she can do as an individual to change the statistics on prematurity and to insure a good start for her own child.

Infants born two weeks or more beyond the expected

date are said by some authorities to be *postmature*. Not everyone agrees with this definition, however; some physicians insist that this period is too brief, and some deny that there is such a thing as postmaturity at all. In any case, the variation among human beings is as great in this respect as in any other, and the period of gestation has occasionally been extended by as much as three weeks without giving any evidence of postmaturity.

It is now possible to predict with a high degree of accuracy which pregnancies will end with a postmature infant. Dr. C. N. Smyth has recently used an oxytocin test: small doses of oxytocin are given to the mother intravenously at one-minute intervals. This substance usually causes the uterus to contract. If the uterus shows a low degree of irritability day after day, the child is likely to be postmature.

This is a useful test, because postmature infants are in slightly greater danger from anoxia and cerebral hemorrhages than children born at term, and if the physician and the mother are alerted to the possibility, they can take the precautions during delivery that will help the child to an uneventful birth. The risk of postmaturity is greatest with the first child of a mother over forty years old; this may be because her entire reproductive system is showing the effects of aging and is less efficient at performing the complex functions of the birth process.

Cesarean section is the name given to the operation by which a child is taken from the uterus by cutting through the walls of the abdomen and the uterus. Julius Caesar was said to have been born this way, and to have given his name to the operation. Some authorities say that this legend is almost certainly untrue, because his mother would certainly have died of it during that period of history, and we know that she lived for a considerable time after her son's birth. Actually, many women did survive this operation in Roman times, and the legend may be quite true. However, a more likely explanation is found in the Roman law of about 725 B.C. known as the *lex caesarea,* in which it was ordered that this operation be performed on any woman dying during the last few weeks of pregnancy, in the hope of saving her child.

About 2 percent of all deliveries are performed by Cesarean section. This is a highly successful procedure, and although it is naturally more dangerous than a normal birth, the dangers are easily identified and guarded against.

When the operation is performed with a spinal anesthetic, the oxygen content of the blood is reduced, and this may result in some degree of anoxia; but unless it is compounded by other difficulties of breathing, the child is not likely to be damaged by it.

We know very little about the effects of Cesarean section on a child's later growth and development. Dr. Marion Kenworthy has said that "the Cesarean sectioned child is prone to be less sensitized—he cries less, is markedly less irritated by the contacts of handling, etc.—than the first-born child delivered through the birth canal." Not all observers agree with this, however. In fact, very little work has been done in this field. A few mothers who have had both Cesarean-delivered and normally delivered children report that the head of the Cesarean-sectioned child is noticeably rounder than the molded head of the child normally born, and his skin is much less wrinkled.

Human beings, like apes and elephants and other animals who have long periods of gestation, are usually born one at a time. We assume that these two facts are related: it is difficult for a number of individuals, all competing for survival, to be maintained together in one uterus for so long a period. When multiple births occur in human beings, the condition of the infants bears out this assumption. There are definite disadvantages in being a member of a multiple birth.

The best calculations, based upon the theory of probability, show that multiple births may be expected to occur in approximately these frequencies:

Twins in 1 out of 87 births
Triplets in 1 out of 7,569 births
Quadruplets in 1 out of 658,507 births
Quintuplets in 1 out of 57,289,761 births
Sextuplets in 1 out of 4,984,209,207 births

There are only about fifty authentic records of quintuplet

births, and of sextuplet births there are less than six authentic records in all of the history of mankind.

The chances of a multiple birth vary with the age of the mother. Adolescent mothers have the fewest—six per 1,000; the rate increases as the age advances until the age group between thirty-five and forty, where there are sixteen multiple births per 1,000 births. After that, the incidence declines. Some observers have suggested that mothers over forty conceive two or more children at a time more often than younger women, but that because of the aging of their reproductive organs, they are unable to maintain all the infants. This causes the weaker ones to die, perhaps early in the pregnancy. In any case, multiple pregnancies occur thirteen times out of each 1,000 in women between forty and forty-four, and only eight times in each 1,000 in mothers of forty-five or over. Negroes have a higher rate of multiple births than whites—fourteen per 1,000 as compared with ten per 1,000.

One of the largest studies of twins was carried out between 1950 and 1956 by Drs. A. F. Guttmacher and S. G. Kohl, who studied some 1,300 twin deliveries in several American hospitals. The median birth weight of these twins was 5¼ pounds, which contrasts strikingly with the average weight of a baby born at term in a single birth—7 pounds. The boys were, as a rule, heavier than the girls. Fraternal twins—that is, twins derived from separate eggs—were heavier than identical twins—that is, twins derived from the same egg.

The members of a multiple birth have greater difficulty in surviving their first year than do single children, and the more individuals in the birth, the greater the difficulty. Twins born to very young mothers and to mothers over forty-five are in the greatest danger, but this is true of single children as well, and is probably a reflection of the immature and aging reproductive organs in these mothers.

It is not surprising that twins, triplets, and quadruplets have a difficult time before birth. The mother's system is designed to nourish and protect only one child; when it is suddenly called upon to double or triple its work, it may partially fail.

For this reason, it is particularly important to spot multiple pregnancies as soon as possible; with good care, the children's chances of survival are greatly increased. Nutrition, especially, is important. We have seen over and over again that nutrition is vitally important in all pregnancies; in multiple pregnancies, the need for a proper diet is multiplied by the number of children.

If it is possible for the mother of twins or triplets to go to a rest home for the last few weeks, or even two months, of her pregnancy, this will help the children more than anything else except food. Admittedly, this is difficult to incorporate into most household routines and most budgets, and often it is not possible at all, but when it is done, everyone involved benefits.

Another important consideration in multiple births is the supply of oxygen. Investigators have found that circulation is slower in twins than in single children, and although there is double the supply of blood, there is not necessarily double the supply of oxygen. For this reason, a woman who is pregnant with twins should be particularly careful about this. She should not smoke. She should certainly have as little anesthetic as possible during delivery. And she should discuss with her physician the possibility of extra oxygen during the delivery.

Since multiple-birth children are more likely to be born prematurely than are single children, they are subjected to hazards that single full-term children never encounter. Some 50 percent of twins weigh less than 5½ pounds, and the average twin pregnancy ends three weeks earlier than the average single pregnancy. Thus the woman pregnant with twins should consider that she is one of the women likely to have premature children and take care of herself accordingly.

After the birth of the children, she will need extra help, too. She has not only double—or triple—the work to do, but the children themselves are likely to be more difficult in their early infancy than single children are. In the interests of her own health and sanity, as well as the health of the children, the mother of newborn twins should recognize this early and make provision for more help than she would need with a single child.

There is still much work to be done in the field of multiple births, particularly in understanding how to insure these children adequate nourishment and oxygen so that their chances for development will be as good as those of their single brothers and sisters. In the meantime, the prospective mother of twins must take special care of herself on behalf of her children.

PHYSICAL INFLUENCES

you think. Also, the physical condition of the expectant mother greatly affects, because being pregnant has so much to do with... even the mental and the emotional life of the whole world-to-be, unborn.

XIII—PHYSICAL INFLUENCES

A MOTHER'S WOMB has always symbolized for mankind complete protection against the harshness of the world. Here, we have believed, an individual lies in a warm bath, enveloped, nourished, and above all, sheltered. Here, no harm can come to him; he is safe. Or so we have believed.

We must now change this symbol. If we are to believe that a child before birth can be benefited by his mother's good diet—and this has been demonstrated over and over again—we must also believe that he can be harmed by her poor diet. And this leads us to believe that he can be both benefited and harmed by a wide variety of other influences. This book, in fact, is devoted to an exploration of what these influences are, and how a mother can take advantage of the beneficial ones and avoid the harmful ones. The symbolic meaning of the womb might now be changed to agree more exactly with established truths: that although the child before birth is more sheltered and protected than he ever will be again, he is never, even here, altogether isolated from the world around him.

Many of the influences that affect this child are indirect. They act on a gland or an organ, which then itself reacts; it is sometimes this reaction, rather than the original action, that reaches the child and affects his growth and development.

Some of the influences, however, are direct, and among the most important and interesting of these are the physical factors in his environment. The mother's activities, for example, act continuously on her child: her breathing, the beating of her heart, her movements when she walks and runs. Also, the physical conditions of the surrounding world—temperature, pressure, noise, vibrations, the pollution in the air, even the season of the year—are now known to act on a child before birth.

The effect of these influences depends to some extent on the child himself. We once believed that a child before birth lies passively, accepting without effort the nourishment that is delivered to him. Now we know that this concept is not precisely correct. The amount of milk an infant drinks is not altogether determined by the breast from which he sucks; likewise, the extent to which he is affected by his environment before birth is partly a result of his own temperament and strength. His reactions are determined by a complex combination of reflexes common to all human beings, particular inherited tendencies, and specific factors in his environment. In addition, each reaction sets the stage slightly differently for the reaction to follow.

In other words, the two-way relationship between an individual and his physical environment, which does so much to determine his life pattern, begins not at his birth, but at his conception.

Every time a pregnant woman breathes, or changes her position, or moves at all, she is varying the pressures on her child. These changing pressures on his skin provide the child with a steady flow of mild stimulations, and some investigators consider these important to him. Drs. L. Ourth and K. B. Brown suggest, in fact, that they may determine a child's sensitiveness to handling after he is born. They add that a child may actually need similar pressures on his skin after birth, because he has grown accustomed to them before birth. His mother can make his adjustment to life easier if she will hold him, and rock him, and caress him, in something like the amount that he enjoyed before birth.

These investigators also point out that children differ widely in this need for mothering. Can this difference be related in any way to differences in prenatal activity? Drs. Ourth and Brown conducted a small-scale study, with two groups of ten normal newborn children, to see whether or not there was any validity to this idea.

The first group of infants was handled no more than

was necessary under routine hospital care. The second group was held and rocked rhythmically at each feeding time. The first group cried significantly more than the second group. The investigators point out that this is not conclusive evidence of anything; much more substantial research is needed before we make any definite statements in this area. But mothers from the beginning of time have known that holding and rocking a small child will comfort him and stop his crying; it seems reasonable to suppose that this is related to the pressures and motions he has known since he was conceived.

Does a child hear sounds before he is born? Some observers have said "No," because the child's Eustachian tubes are flooded with amniotic fluid, but most of the evidence points to "Yes." From about the twenty-eighth week of pregnancy, a child will respond actively to sounds of all kinds.

For example: A woman who worked as a stenographer reported that she was forced to give up her job during the eighth month of pregnancy because every time the typewriter carriage was returned to begin a new line, the sound caused her child to jump. With thirty lines to a page, this activity was more than she could take with comfort.

Other mothers have reported that their children became extremely active at concerts of orchestral music, to the point where these women were forced to leave the concert hall because of the discomfort. Loud piano playing close by has the same effect. This suggests that an enterprising student of these matters might try to determine what differences in effect would be produced by the music of different composers: would a child before birth react differently to Schönberg, for example, than to Mozart?

The vibrations caused by a washing machine are also capable of causing a child to become extremely active, and so is tapping on the side of a bathtub while the mother is lying in it. During episodes of this kind, the heartbeat of the child increases by ten beats per minute, or more, and the older the child, the greater the increase in activity and the change in heartbeats per minute. This is particularly interesting, because normally the heartbeat rate tends to decrease as the child grows older.

When a sharp, loud sound is made close to the mother's body, near the child's head, he may move so suddenly and convulsively that the movement can be called a startle reflex. This is a complex response involving the muscles and organs, but it is also connected with fright, or alarm, or surprise, or some similar emotional state. Can we say, then, that a child before birth is capable of fright, or surprise? We cannot say this yet, but at the same time we cannot say that he is *not* capable of these emotional responses.

We *can* say that a child is able to learn before he is born; this was demonstrated by Dr. D. K. Spelt, who used sounds, vibrations, and pressures in an experiment. This was truly a historic demonstration, conducted in 1948, with the assistance of sixteen pregnant women in the last two months of pregnancy, and their children. Dr. Spelt used an automatic clapper to make a loud noise and an ordinary doorbell with the clapper removed so that it would produce only vibrations.

The demonstration began when Dr. Spelt set off the clapper near the women and recorded the reactions of their children. The activity was considerable. Then he set off the vibrator as close as possible to each child. No response at all. The next step was to set off the clapper and the vibrator at the same time. This was done fifteen to twenty times. After this, the vibrator was set off alone; this time the children responded, as they had not done previously. Furthermore, they continued to respond to the vibrator alone as many as eleven times before the response died away.

This is a classic example of a conditioned reflex. The fetuses were actually taught to do something they had not previously done, by connecting it with something they had previously done, and then removing the original stimulus. Other investigators have questioned this demonstration, pointing out that very young infants can be conditioned only with the greatest difficulty. The answer to this is simply that Dr. Spelt succeeded in getting genuine conditioned responses from these sixteen children.

The significance of this demonstration, and its implications for the future, are still uncertain. Many more studies are needed before this question can be settled satisfactorily.

The noises of bombing and shelling during wartime are among the more serious of these influences on children before birth, but it is difficult to sort out the effects of the noise from those of all the other aspects of war. Dr. Lester Sontag, who wrote one of the few papers on the consequences of war on children before birth, suggested that the sound of bombing would produce the same kind of neurosis in a child before birth as is produced by highly disturbed emotions in his mother. In fact, as he pointed out, during a bombing the mother's emotions would surely be disturbed in any case, and these two factors would be inseparable. The specific results in a child's behavior would appear at birth, when he might have great difficulty with feeding and digestion.

Industrial noises, too, are probably capable of affecting a child. No work has been done directly on this problem, and we cannot be certain of the amount or kind of influence such noises would have, but this has been fore-shadowed by a study conducted in the laboratory. In 1961, Drs. H. Ishii and K. Yokobori, working with several groups of mice, demonstrated that in the case of these animals, noise does have a connection with the condition of the developing animals. The investigators subjected each group of animals to different degrees of noise for several days. At the conclusion of the demonstration, the newborn animals that had been in the quietest environment were the largest and the healthiest and had among their number the fewest premature births. The proportion of poorly developed animals and premature births increased directly with the increase of noise.

Has this study any implications for human beings? There is no way of knowing at present. But if we consider the effect that loud continuous noise has on the nervous systems of healthy adult human beings, we may agree that such a noise is quite likely to affect in some way a human being who is only just developing.

Not all sounds are harmful to a child before birth. There is one sound in particular he hears continuously throughout his entire prenatal life, which is probably beneficial to him: his mother's heartbeat. Dr. Lee Salk has suggested that this steady rhythm may be in part responsible for the lack of tension that characterizes a child before birth. And the universal attraction of music may also be related to the mother's heartbeat. Music is sound organized around a regular beat, and this may appeal to us through the deepest memories we have, from our prenatal days. Dr. Salk has observed, too, that by far the greater number of women hold their infants on the left side, where the children can more easily feel their mothers' heartbeat. Even left-handed mothers, who might be expected to hold their children on their right arms, to free the left, hold them on the left side more often than not. Why is this? We do not know the answer, but the suggestion has been made that mothers recognize subconsciously that a child is quieter when he can hear his mother's heart.

This suggestion was the basis of an interesting demonstration by Dr. Salk, which he described in a paper entitled "The Effects of the Normal Heartbeat Sound on the Behavior of the Newborn Infant." He showed conclusively that when a newborn child can hear the sound of a normal heartbeat, he is quieter, he is healthier, and he gains weight faster than when he cannot.

Dr. Salk set up this demonstration in the nursery of Elmhurst Hospital, New York, with two loudspeaker systems, through which he played authentic normal heartbeat sounds at the rate of seventy-two paired beats per minute. The sound was kept at a moderate level, and it was continued day and night throughout the period of the demonstration. The infants observed were the regular inmates of the nursery, who were placed there immediately after birth and kept there for four days, except for their feedings, when they were taken to their mothers. The demonstration continued over a period of sixteen weeks: the heartbeat sound was played for four weeks, turned off for four weeks, played for another four weeks, and turned off for the last four weeks. During the entire sixteen-week period, the weight of the children was closely watched, and Dr. Salk and his assistants kept records of the amount of crying at all times.

The most startling observation was concerned with the weights of the children. The heartbeat sound was clearly beneficial from the point of view of their gain in weight. Sixty-nine percent of the children who had listened to the heartbeat gained weight in their four days in the nursery;

only 33 percent of the others gained. Furthermore, the average amount of gain in the first group was forty grams, a considerable amount for children of this age. The group of infants who had not heard the heartbeat not only did not gain weight; they showed an average *loss* of weight of 20 grams.

The children who heard the heartbeat also cried less than the others. During the period when the sound was being played, one or more of the babies was crying 39 percent of the time; when it was not being played, one or more of them was crying 55 percent of the time.

Also, Dr. Salk noticed that those children who were in the nursery while the sound was being played breathed more deeply and more regularly than the others. There seems to be no question that the sound was calming to the newborn children.

Inadvertently, the investigators were able to see what happened when the quality of the sound changed. The tape became worn during the course of the demonstration and began to make a hissing sound; there was a noticeable increase in crying and restlessness among the children. On another occasion, the heartbeat rate was increased from 72 to 128 beats per minute; the result was the same.

The gain in weight among the children who heard the heartbeat was probably a result of their less frequent crying, according to Dr. Salk. Their energies were used to grow, whereas the energies of the other children were used to cry. The implications for the general health of children is obvious: we have seen over and over again that larger children are stronger and more resistant to infection and stresses than smaller children. When we reduce the amount of crying, we may at the same time improve their health.

Dr. Salk used this same technique with older children who had been having sleeping difficulties. Some of these children were so tense that they lay awake for several hours every night before they could relax and go to sleep. These children in particular were greatly helped by the playing of the heartbeat sound nearby. In fact, some of them went to sleep in fifteen minutes.

The sound of a normal heartbeat during the early weeks of a child's postnatal life might well contribute to a better emotional development later in his life, Dr. Salk believes. He says about older children: "Sleeping difficulties in children are not uncommon and most often are the result of a fear of losing contact with the parents, which to the child means abandonment." But the sound of the heartbeat, which to the child is connected with the greatest security he has ever known, may succeed in calming his fears where all else has failed. It is very interesting, in this connection, to note that when a woman pediatrician allowed her five-year-old son to listen to her heartbeat through a stethoscope, he remarked, "Mummy, I can hear your love flowing to me."

A child before birth is also acutely aware of any changes in the temperature and pressure of his surroundings. Normally, he lives an aquatic life, bathed in the amniotic fluid, and the temperature and pressure of his immediate surroundings are unchanging. Sometimes, however, these conditions must be altered, and then we can see to what extent he notices them and how he reacts to them.

One of the recent developments in the technique of surgery is to cool the body of the patient. This successfully combats the damaging effects to the brain of large amounts of anesthetics. When the body is colder than normal, metabolism is retarded, circulation is slowed, and as a result the brain needs much less oxygen than it requires at normal temperatures. This is particularly valuable when the surgery is protracted and the amounts of anesthetic might become dangerously large.

Drs. F. Wilson and C. B. Sedzimir have reported the experience of a young woman, thirty-two weeks pregnant, who was to have an operation on her brain: her body was cooled over a period of several days by placing ice packs around her. As her temperature dropped, the heartbeat of her child dropped, too, from 160 to 85 beats per minute. Toward the end of the operation, she was given an injection of trimethaphan, a drug that acts upon the nervous system and produces a relatively bloodless field for the operation. Within five minutes, the child began to kick violently. The physicians suggested that the kicking may have been due to anoxia in the child brought on by the drug. Fortunately, it was not severe enough to harm him; he was born normal, and at the time of the report he was thriving.

The temperature of another patient, twenty-four weeks pregnant, was reduced to 86° Fahrenheit before an operation. The heartbeat of this child fell from 180 beats per minute to 120. This, too, was a successful operation, and the child was born at term, healthy and normal.

Sometimes the pressure in the amniotic sac may be reduced or increased, when for some reason the sac must be punctured, and although this strikes most laymen as a dangerous operation, actually it seems to be safe. An operation called transabdominal amniocentesis is sometimes performed on a pregnant woman to determine the sex of her child: a syringe needle is used to puncture the abdominal wall and the amniotic sac and withdraw amniotic fluid with bits of skin shed by the child. By studying the cells of these bits of skin, investigators have been able to determine the sex of the child as early as eight weeks after conception. Fifty women who had this operation had no difficulties as a result of it, and apparently none of their children suffered from it.

The same operation is also performed to relieve the condition known as hydramnios, an excess of fluid in the amniotic sac. Dr. L. C. Rivett has reported on fifty women who had this operation: again, there were no harmful results to either the mothers or the children.

Demonstrations of this operation using laboratory animals, however, suggest a general precaution: it would be unwise to perform this on a human being earlier than the tenth week of pregnancy. In the laboratory demonstrations, the development of the young animals was definitely interfered with when the operation was performed early in pregnancy.

In general, surgical operations on a pregnant woman do not threaten her child's welfare. The primary danger to him arises from the analgesics and anesthetics given to his mother. These, as we saw in the discussion of oxygen starvation in Chapter IV, rob the mother's bloodstream of oxygen, which in turn results in a lack of oxygen in the child's bloodstream. And this interferes with his development. If this danger is recognized and dealt with —for example, by using spinal instead of general anesthetics—the child is apparently in no particular danger from surgery. The one exception to this is open-heart surgery. Only two cases of this have been reported when the patient was pregnant, but in these two, the children were lost.

Various other kinds of pressure act upon a child before birth. During labor, for example, the child is normally subjected to considerable pressure, and as we shall see in Chapter XV, this can sometimes result in a slow heartbeat and even some degree of anoxia.

An interesting device for lessening the effects of this pressure has been worked out and demonstrated by Drs. O. S. Heyns, J. M. Samson, and J. A. C. Graham. The device consists of a plastic suit, fitted with a powerful suction pump. Over three hundred women have used this suit-pump combination during labor. They wore the suit and worked the pump whenever a labor pain started. The effect of this was to lower the pressure on the child within the uterus by lowering the outside pressure against which he struggled. This appears to be a more complicated and difficult arrangement than could ever be made generally available, but it appears to have had definite advantages. Later, all the mothers who had had other children told the physicians that the child born with the help of this device was noticeably more "advanced" than the other children. The investigators believe that the lower pressure served to increase the amount of oxygen available to these children; this meant that none of them suffered even slightly from anoxia and brain damage during the last part of pregnancy and at birth.

Intercourse during pregnancy may also be an occasion for undue pressure on the child. Although there is some disagreement among physicians on this subject, many gynecologists now recommend that intercourse be continued during pregnancy, through the seventh or eighth month, but that particular care be taken to avoid positions that would involve pressure on the woman's abdomen. The risk of infection, also, must be taken into account, but with scrupulous cleanliness this risk can be minimized.

Some societies in human history have forbidden sexual intercourse during the entire period of pregnancy; the Talmud and the Koran are explicit on this point. Chinese culture adopted the same attitude. Modern scientific objections to sexual relations during pregnancy have centered on the dangers of miscarriage and infection.

For psychological reasons, however, a continuation of sexual relations during most of pregnancy is believed by present-day authorities to be important to both husband and wife. In fact, to deprive them of this during the entire period is perhaps even to endanger the marriage. If both partners understand that special attention must be given to avoiding miscarriage and infection, there is no sound reason why they should not continue. Gentleness, the use of lubricants, cleanliness, and the avoidance of the flexed position, in which the woman's knees are drawn up almost to the breast, will reduce the physical risks to both mother and child.

Stimulation of the nipples in pregnant or nursing women causes marked contraction of the uterus. After a child is born, when he is nursing vigorously, this stimulation serves a useful purpose in promoting the quick return of the uterus to its normal size. Before birth, however, contractions of this kind may be harmful to the child. They may cause a decrease in the oxygen supply and result in damage to him. So this stimulation or caressing of the nipples should be avoided, particularly late in pregnancy.

Another example of possible pressure on a child before birth is the situation that occasionally exists when twins or triplets interfere with each other's freedom of movement. When a child is unable to move a particular joint, it sometimes becomes rigidly fixed. Dr. Denis Browne and others have shown that this is a common cause of clubfoot and other limb deformities.

Low atmospheric pressure, such as one encounters in the high mountain regions of the earth, produces a certain degree of anoxia in all the human beings who live there, and in this way it affects the development of a child before birth. In the Andes, for example, a child runs a greater risk of cardiovascular disturbances than does a child at sea level, owing to the reduced amount of oxygen in the air. For this reason, mountain climbing is not recommended during pregnancy, not because of the muscular activity involved—which would not harm a woman accustomed to it—but because of the low oxygen content of the atmosphere, which is not good for the child. The low pressure, as such, has no effect on the child;

in fact, the atmospheric pressure inside the uterus, in which the child lives for nine months, measures 35 to 40 millimeters of mercury, which is even lower than that at the summit of Mt. Everest. Nor is he affected by modern air travel, particularly in pressurized planes.

Electric shocks of 220 volts sustained by pregnant women have been reported in the medical journals only four times. In each case, the child was so severely affected that he died, either immediately or soon after. Apparently, a child before birth is unusually vulnerable to electric shock.

Electroshock therapy, on the other hand, has evidently not harmed the few children whose experiences have been reported. This therapy is sometimes used in the treatment of emotional disturbances. Dr. D. E. Sobel observed a total of eighty-two women who received shock treatment during pregnancy. He saw all their children at birth and followed up thirty-five of them several months later. He found no children who appeared to have been affected by the shocks. According to studies of various laboratory animals, Dr. Sobel might have expected to find that the children of women who were treated very early in pregnancy were harmed by the shocks; this appears to be true in the case of demonstrations in the laboratory. Dr. Sobel's group of women included five who were treated during the first eight weeks of pregnancy, none of whose children were affected at all, but as he points out, this is too small a group from which to draw any final conclusions.

Do the seasonal changes of the earth have any influence on the growth and development of children before birth? Which is the stronger man, the one born in summer or the one born in winter? Which is the more intelligent? Which will be the leader and which the follower?

This is one of the fascinating questions that have occupied man's thoughts at intervals since he began thinking about such matters at all. Astrology is based on this, and the large number of people who believe the teachings of astrology proves how eager men are for answers to these questions. Fortunately, more scientific studies are now being carried out, and the conclusions our scientists are reaching are quite as interesting as any astrological mumbo-jumbo could possibly be.

In general, investigators in this field agree that children conceived during the autumn and winter period tend to be more favored than those conceived during the spring and summer. Statistically, they are heavier, they are healthier, they have higher I.Q.'s, they go to college in greater number, they are more frequently included in *Who's Who*. Conversely, more mentally deficient individuals are conceived in the spring and summer months than in the autumn and winter. Moreover, complications of pregnancy are more frequent among mothers of children conceived in the warmer months.

At first glance, these conclusions appear to be as illogical as astrology. Further studies, however, have explained them, and they bring us full circle once again to the subject that crops up in nearly every context: nutrition.

Nutrition, it appears, is the foundation for these differences based on season of birth. The explanation is simple and reasonable enough when it is pointed out: in hot weather, all of us eat less than in cold weather. Pregnant women, in particular, who may be uncomfortable anyway, and are trying to keep their weight down, are more likely than the rest of us to skimp on their diets in very hot weather. The immediate result of this is to slow down the growth and development of their children. And most seriously affected of all are the children who happen to be at that period of their development when the brain is being formed. This period is between the eighth and twelfth weeks of pregnancy.

Therefore, when very hot weather happens to coincide with this period of brain development in an individual child, the prospective mother must be even more than normally careful about her diet, to be sure that she includes in it the proper amount of protein, even though she might rather live on salad and fruit. Meat, poultry, fish, and eggs, in one or more daily servings, will supply the proper amount of protein.

Dr. Hilda Knobloch and Dr. Benjamin Pasamanick made this interesting correlation between the development of the brain and the season of the year in a study of the retarded children and adults in the Columbus, Ohio, State School. The age range of the subjects of the study was particularly wide—some were born as early as 1913 and others as late as 1949.

The investigators reasoned this way: if hot weather does influence the diet of pregnant women enough to affect the brain development of their children, the individuals who were in their eighth to twelfth week of prenatal life during the hottest months of the year would be the most seriously affected. The study proved this reasoning to be accurate.

Furthermore, the hotter the summer, the more defective children were born the following winter and spring. In fact, during the cooler summers, there was actually no increase in the number of retarded children whose eighth and twelfth weeks coincided with the summer months.

Other investigators agree with the conclusions of this Columbus study. Drs. R. Pintner and G. Forlano analyzed sixteen large-scale studies on the relationship between season of birth and intelligence and found that in fifteen of them the I.Q. was lowest among those children who were conceived in the spring and summer. In only one study was the lowest I.Q. found among children conceived in autumn and winter; this study included 337 children, a small number compared with the numbers included in the other studies.

A light diet is actually beneficial during hot weather during most of our lives; during pregnancy, it may be harmful to the child. Pregnant women must take particular care, especially during hot weather, to eat well and to include plenty of protein, which is the most important single item of the diet for growth.

The increased pollution of the air we breathe is a serious matter for all of us, but to a prospective mother it is of particular concern; she also has her child to consider and protect. And the greater the degree of pollution of the air in her own lungs, the less oxygen there will be in her blood, and the less she will be able to pass on to her child. An industrial, mechanized society such as ours threatens our oxygen supply on all sides, and in our cities in particular air pollution is a major source of danger. Dr. Frank L. Rosen recently described the air we breathe as "a giant open sewer." He went on to say, "Since we have no choice but to breathe it, it is high time we paid some attention to the garbage we spew into it."

A prospective mother can hardly improve the air pollution ordinances of her community in time to help her own developing child, but she should be aware of the major dangers and avoid those she can. Also, by becoming aware of them, she may be in a better position in the future to act responsibly as a citizen and as a protector of future generations.

Gasoline engine exhaust, for example, contains several substances that rob the blood of oxygen. Carbon monoxide we have already discussed, on pages 112–113. On the streets of a large city, or near a major highway, these substances are concentrated particularly heavily. A pregnant woman should stay away from these concentrations if she can.

Another big-city contaminant is benzopyrene. A study of this substance by the U. S. Department of Health, Education and Welfare gives the amounts in the atmosphere in several kinds of environments, ranging from a state forest, where there is almost none, to a city street, where the amount is considerable. Dr. Rosen calculates that a city dweller, living in the midst of this concentration of benzopyrene, would inhale an amount of it equivalent to the amount in two packages of cigarettes daily.

Factories, chemical plants, and airports are other sources of air pollution. A single jet airplane taking off, for example, pollutes the air with the same amount of exhaust as would come from 6,850 automobile engines idling for one minute each. And the smoke pouring from the stacks of busy industrial centers indicates a prosperous economy, but it is definitely not good for the oxygen content of our blood.

Pesticides are still another threat to the purity of the air we breathe. Two major indictments have been made recently against our indiscriminate use of these chemicals. Rachel Carson's book *Silent Spring* is a cogently written report on the widespread use of poisons to control animal and plant life, and the dangers that these poisons offer to human beings. The other was made by the President's Science Advisory Committee, which reported in May, 1963, that pesticides pose a serious hazard to human life and health and recommended stronger controls of their use.

Thanks to these two reports, there now seems to be

no question that as a society we will try to improve this serious situation on a nation-wide official scale. However, no amount of government supervision and regulation will instill common sense and restraint into individual users of pesticides and other chemicals. Individuals, as always, will have to supply their own common sense. And pregnant women, in particular, must recognize the dangers to their children before birth and take particular precautions to protect them.

The figures show the extent of the danger. Accidental acute poisoning in human beings has been caused by about fifty different pesticides, according to the report from the President's Committee, and approximately 150 deaths occur every year in the United States from this mishandling. About half of these are children, who are accidentally poisoned in their own homes. The number of poisonings that are less than fatal is impossible to state; but a special committee in California, where about 20 percent of our pesticides are used, estimated that about 3,000 children take various amounts of these chemicals into their systems every year.

Children before birth are as susceptible as the rest of the population to poisoning. The oxygen content of their blood is affected, as we have seen again and again, by any substance that travels in the mother's bloodstream. Furthermore, the chemicals themselves appear to be capable of crossing the placenta and entering the child's body. At least, laboratory demonstrations have shown that this is true in animals. Residues of insecticides have been recovered from the milk of nursing mothers.

The exact effect of these many pesticides on a developing human being is not yet known, but since they are poisons, each especially developed to kill or upset the cellular metabolism of some living organism, it is reasonable to suppose that they may have a damaging effect on a child before birth.

The solution is not, of course, to stop using pesticides altogether. They have been among our most useful tools in raising the standards of health and comfort all over the world. Many contagious diseases—malaria, typhus, yellow fever, and others—that once were world-wide scourges can now be controlled or even eradicated by using pesticides. The production of food has been enormously increased, in quantity and quality, through the use of pesticides. Because of these chemicals, millions of human beings are today alive and healthy who would have died of disease or suffered from malnutrition without them.

Viewed in this light, perhaps 150 deaths per year from the misuse of pesticides seems a small price to pay. But of course we cannot allow that either. And we need not allow it. With an informed public and proper regulations to control the large-scale use of these chemicals, we can have the benefits from them without damaging ourselves.

As it is, we all inhale quantities of pesticides. And the more conscientious we are in spraying our houses and gardens to rid them of flies, moths, mosquitoes, and the various threats to our vegetables and flowers, the more poison we inhale into our own lungs. Large-scale sprayings —of orchards, of town dumps, in community-wide fly-control projects, and the like—present the individual with difficulties. A pregnant woman who is eager to protect her child can hardly leave town every time a nearby apple-grower sprays his orchard, even if she knows when he plans to do it. But she can investigate the local ordinances, to see whether they embody the most recent information on this subject and place sufficient controls on the local sprayers. And she can surely control the amount of pesticide that is sprayed around her own home.

The poisons in pesticides enter our systems in other ways, too. They are in the food we eat, the water we drink, in the moth-proofing of our rugs and blankets, and they may even contaminate the surfaces of dishes and cooking utensils in our kitchens.

According to the President's Committee, physicians are generally not aware of this threat to our health. The situation has grown in the dark, as it were, shielded by the unquestioned value of pesticides to all of us. It is only recently that their dangers have become clear, and many busy physicians have not yet noticed quite how widespread these chemicals are, nor how poisonous they are turning out to be to human beings, nor specifically how they are affecting our health. Ordinarily, we might relax and allow this lag, knowing that our doctors would catch up soon with the work in the laboratories. In this case, however, and particularly where pregnant women and their children are concerned, we cannot be so casual. Every pregnant woman should read Rachel Carson's book and the report of the President's Science Advisory Committee, and attempt to keep up with the information that appears frequently in newspapers and magazines.

These studies on the various physical factors that are known to influence the growth and development of a child before birth are surely only the beginning of our knowledge in this field. Even now, the ways in which we know that a child can be affected by his physical surroundings are varied and important: indirectly, through his nutrition and the oxygen in his blood; and directly, through sounds that comfort him and sounds that startle him, through temperatures and pressures both low and high, through changes in temperatures and pressures, through electrical shocks, through contamination of the air, and through the food and the water that enters his body. The list will soon be longer.

The duty of the mother of a developing child is to keep abreast of these additions to our knowledge, to avoid where she can the influences that may harm him, and to take advantage where she can of those that can benefit him.

Specifically:

A pregnant woman may help her child if she avoids exposing him to continued loud noises.

She may help him if the air she breathes is not polluted, and if she controls as best she can the pesticides that are used in her vicinity.

She may help him if she continues to eat well even during hot weather, particularly if her child is still in the early stages of his development during the summer.

She may help him if she is careful about cleanliness and gentleness when she has sexual relations.

XIV—RADIATION, X RAYS, AND FALLOUT

RADIATION IS nothing new to human beings. Whether we knew it or not—and until relatively recently we did not know it—human beings have always been exposed to radiation, from the sun, from outer space, from rocks and soil, and even from the food that was grown in that soil. And to some extent we have adapted ourselves to that amount.

What frightens us now about radiation is that we are threatened with enormously increased doses. This threat has hung over us only since we learned to release the energy in the atom. We are not so adaptable that we can absorb unharmed the energy, in the form of radiation, that we are now capable of unleashing. The question of nuclear war is frightening enough, with the possible destruction of large parts of the civilized world and the sufferings of the survivors. In addition, we are faced with the present practical question of the effects of bomb testing. What influence does the fallout from those tests really have on the human population of the earth? In particular, how can fallout affect children before birth? This is a subject that every human being, and especially every prospective mother, should try to understand.

Scientists have been focusing their attention in recent years on this question, and their conclusions are illuminating. In many areas, details are uncertain and experts disagree with one another, but the general outlines of the prenatal effects of radiation have been established and are perfectly clear.

The bare fact is this: radiation can be seriously damaging to a child before birth.

Massive exposure to radiation is, of course, damaging to all living things, and lower degrees of radiation, if they are continued over long periods, can also be harmful. Even X rays, one of the most useful tools of medicine, may be dangerous to a child before birth—pregnant women should be particularly careful, even about dental X rays. For unless the proper precautions are taken these rays will reach the mother's ovaries and the child. X rays scatter unless they are screened.

A child's development can be interfered with by radiation. He may be rendered sterile. And most far-reaching of all, radiation can cause permanent changes within the genes of individual cells. These changes do not die with the individual; they can be passed on to future generations. They are called mutations, and 99 percent of them are harmful.

The mechanism of the influence of radiation on human life is not simple; it has its base in the nature of matter itself.

All matter is made up of atoms. Each of these atoms is a miniature solar system: a nucleus in place of the sun, surrounded by a swarm of electrons spinning in orbit.

The nucleus is made up of particles known as protons and neutrons: the protons carry a positive electrical charge; the neutrons have no charge. The spinning electrons are light, fast-moving particles that are negatively charged. The electrons are maintained in orbit around the nucleus by this difference in electrical charge; the negative electrons are attracted to the positive protons.

Each element in nature is composed of a particular kind of atom, with a definite number and arrangement of protons, neutrons, and electrons. Each element is different in this respect from every other element.

When atoms come together with other atoms, they combine in orderly fashion with each other to form various orderly, functioning compounds. For example: One of the most familiar and simple of the compounds is water. This is made up of atoms of oxygen and atoms of hydrogen. These two elements, quite different from each other, combine easily in the ratio of two hydrogen atoms to one oxygen atom. This is what is meant by H_2O. It is a definite, well-understood material of our universe. We know what it will do and how to treat it. At certain temperatures, it is a liquid. Above a fixed temperature, at certain pressures, it will turn to a gas, known as steam. Below a fixed temperature, at certain pressures, it will freeze. This is a typical chemical compound, with definite physical characteristics and made up of a definite number and arrangement of basic elements. Thus is our whole material world made up, atom by atom, in all its simplicity and variety and order.

But suppose the internal structure of one of these atoms is disturbed, what then? Suppose a single electron is somehow removed from the solar system around the nucleus? What happens?

First, of course, the fundamental nature of that atom is changed; it becomes something quite different. Its electrical balance is disturbed; it is no longer an atom of oxygen, or hydrogen, or whatever; it becomes unbalanced and unstable.

This means, then, that its combination with other atoms is disturbed. It will not combine with them in the orderly, functioning fashion it did when all its electrons were present. It may combine at random with whatever substance will accept it. In any case, the chances are great that the new combination will not be an improvement, but an aberration.

This kind of disturbance is a result of radiation. For example: A ray of extremely high energy penetrates the solar system of an atom and knocks away some of the electrons from the magnetic pull of the nucleus. The atom then becomes another substance, electrically charged, unstable. This substance is called an *ion*.

The electrons that have been detached then fly outward at random until they encounter another atom. Here they repeat the process. This is called *ionization*.

The high-energy rays are what we mean when we speak of radiation. The changes in the atoms are the specific results of radiation. It is this process that holds danger for the human species.

Sunburn is an effect of solar radiation with which we are all familiar. The rays of the sun can be harmful, as anyone knows who has ever had too much. At the very least they can cause discomfort, even illness. Cancer of the skin can be a result of sunburn; this is an example of the extent to which radiation can change the structure of human cells.

Of all the systems in the human body, by far the most sensitive to the effect of radiation is the reproductive system. The reproductive cells, in the testes of the male and the ovaries of the female, carry the hereditary material from one generation of human beings to the next. Any radiation that reaches these cells can cause changes in them that will then be passed on to future human beings. This is the most serious threat from radiation.

The most pervasive form of radiation comes to us from outer space in the form of cosmic rays. Another form comes from rocks and soil and the food grown in the soil. In addition, the elements radium, uranium, and thorium are naturally radioactive; the nucleus of these elements erupts spontaneously from time to time and emits particles and rays. These natural sources supply over half of the radiation to which we are exposed.

A large part of the other half—about 40 percent of the total in the United States—comes from medical and dental X rays. These are produced artificially in special electrical machines, where a stream of fast-moving electrons is directed at a metal target. This forces some of the atoms in the target to lose their electrons, which are then emitted, in the form of X rays.

Fallout, which consists of radioactive atoms of various materials released by an atomic explosion, is the source of 1.3 percent of our radiation at the present time. We also receive radiation from illuminated dials on wristwatches, pocket watches, clocks, dashboards, and other instruments.

We have learned to measure the amount of radiation by measuring the number of ions formed. The unit of measurement is known as the roentgen, usually abbreviated "r." In the human body, 1 r of radiation causes the formation of one hundred million billion ions. This corresponds to the formation of an average of 10,000 ions per cell. This figure is less alarming than it may appear at first; a single cell consists of approximately 70,000,000,-000,000 atoms (70 thousand billion). Thus the number of atoms in each cell converted to ions by 1 r of radiation is only 0.000000014 percent of the atoms in each cell.

An authoritative estimate has been made of the average total accumulated exposure to radiation that each individual receives during the first thirty years of his life. This period was chosen because in the United States the average age of parents at the birth of all their children is just under thirty years; therefore, the amount of radiation they receive by this age is the amount that could conceivably affect their descendants through their reproductive system. The total dose was estimated at 7.4 *r*. Cosmic rays, naturally radioactive substances, rocks and soil, account for 4.3 *r*, except at high altitudes, where greater exposure to cosmic rays increases the figure to 5.5 *r*. Medical and dental X rays were estimated at 3.0 *r*. Fallout was judged to be 0.1 *r*.

Relatively speaking, this amount of radiation from fallout may seem to be very little, but in fact when even 0.1 *r* is distributed over a world population of 3 billion, it adds up to an enormous amount. The mutations that may result from this world-wide dose concern us all seriously. Dr. James F. Crow, geneticist of the University of Wisconsin, has estimated that this exposure will result in at least 8,000 children in the first generation born with gross physical or mental defects, and a total of 80,000 in future generations. He also estimated that in the next generation radiation from fallout would cause about 40,-000 deaths of children before and immediately after birth, and a total of 700,000 in the long run. Furthermore, he warned that these figures are probably low.

Damage to human beings from X rays is on a different scale altogether. The world population is not exposed wholesale to X rays as it is to fallout. On the other hand, individuals in medically advanced countries receive a much larger dose of radiation from X rays than from any other man-made source.

Human cells are far more sensitive to X rays than was formerly believed. Dr. Theodore T. Puck of the University of Colorado has recently shown this; a dose as small as 50 *r* is fatal for the reproduction of human cells. Since life itself is dependent upon the ability of cells to reproduce themselves, a relatively small dose of X rays over the whole body can be fatal. A much smaller amount is capable of damaging the genes, the basic hereditary materials of the human species. The tragedy in such cases may lie not in death but in survival with seriously damaged equipment.

The influences on children before birth discussed thus far in this book are, with the exception of radiation, influences that affect the child alone, or at most the child together with his mother. The psychological and emotional effects may spread beyond them, but the actual physical harm stops with them. In the case of radiation and its effects on the genes, this is not the case: when the genes within a human cell are affected, the traits that are governed by those damaged genes are transmitted to the next generation and to all the generations that follow. This kind of sudden, well-marked change in any species of living organism is known as mutation.

Mutations do occur spontaneously in all living things; they are now believed to be caused principally by cosmic rays. The rate of this mutation in mankind is not known, but in the lifetime of every individual a certain number of mutations occur as a result of these natural causes. Scientists have guessed that the number may range from one to four. If they occur during reproductive life, they will be transmitted to the children.

Many of these changes are too small to detect. Some, however, are all too evident: mental deficiency, epilepsy, malformations, neuromuscular disorders, and defects of the gastrointestinal and genitourinary systems. About half the children who suffer from these defects—in other words, about 2,000,000 children—are the victims of congenital damage. If the population of the United States were to receive a double exposure to radiation, the number of genetic defectives among us would increase by about 200,000 in the first generation and would eventually be doubled.

Dr. W. L. Russell of the Oak Ridge Laboratory has shown that 1 r of radiation produces in a mouse about one mutation in 4,000,000 genes. This rate can be used to help our scientists estimate the effect in man. We already know the spontaneous mutation rate for certain genes—for the gene that carries the hemophilia trait, for example—and putting these facts together, experts have estimated that a dose of some 30 r to 60 r would double the natural mutation rate in man.

Our mutation rate in the United States is already being increased by our ever-increasing use of medical and dental X rays. These constitute our largest exposure by far to man-made radiation. Furthermore, we have been guilty of an enormous amount of totally unnecessary exposure of our pregnant women and their children to direct radiation.

As long ago as 1924, we began seeing evidence that children may be seriously damaged before birth by exposures to X rays. During the five years following that first indication, nearly 100 children whose mothers had been given X-ray treatments during pregnancy were studied and reported on. Seventy-five of these mothers had been treated with X rays in their pelvic regions; the children, therefore, had been directly in the path of the radiation. Dr. Douglas Murphy, who conducted this survey, concluded that nearly thirty of these children were seriously damaged at birth as an almost certain result of the irradiation of their mothers. He pointed out, too, that in some of the cases the irradiation occurred late in pregnancy, so the danger is apparently not confined to the early months.

Dr. Murphy recommended strongly that a pregnant woman should *never* submit to radiotherapy in the region of her uterus. Moreover, if she should have to have treatment and then later discover that she was pregnant at the time, Dr. Murphy recommended that the pregnancy should be terminated as soon as possible.

Occasional X rays for diagnostic purposes are not dangerous, as Dr. P. W. Toombs, in 1930, was the first to show, after he conducted a survey of the studies described in the medical journals. X-ray therapy requires much larger doses than X ray for diagnosis, and Dr. Toombs found that the therapeutic X rays frequently caused abortions, when the exposure occurred during the early months of pregnancy, and various interruptions of the child's development when it occurred during the later months.

An important book, *X-ray and Radium Injuries,* by Drs. H. A. Colwell and S. Russ, published in 1934, set out the facts as they were understood at the time. The authors concluded, "The case against irradiation of the fetus *in utero* is agreed on by all hands."

Two years later, other voices were added to the chorus

when Dr. J. E. Miller and his co-workers concluded, at the end of their inquiries, that they considered it "advisable to interrupt any pregnancy which has been subjected to therapeutic radiation."

These opinions, all based on solid evidence spread out in medical books and journals for all to see, have been on the record for well over a generation. Yet the practice of exposing pregnant women to X rays has not yet been abandoned. A distinction must be made between X rays used for diagnostic purposes, and those used for therapeutic purposes. The latter are always of much higher dosage, and therefore more dangerous.

Studies of human beings have been supplemented over the years with work in the laboratory. It has been possible actually to produce particular malformations in laboratory animals by giving specific doses of radiation at particular periods during gestation.

This work has greatly increased our understanding of the effects of radiation on human beings. In connection with its effect on intelligence, for example, we could not have had any reliable evidence at all from pregnant women and their children, but when laboratory animals are observed, it is possible to see quite clearly that intelligence is affected by radiation. Animals that have not been irradiated are significantly quicker to learn than those that have, and the slowness to learn is directly connected with the degree of irradiation: the greater the dose, the greater the impairment.

With respect to malformations, in both animals and human beings, the so-called critical period is when the major organs are being formed, the period of organogenesis. We have encountered this period before in our discussions of prenatal influences; it is a hazardous time in the life of a human being. The period extends from approximately the end of the second week to the end of the seventh week of life.

Drs. W. L. and L. B. Russell write about this subject:

For practical purposes, the entire period of organogenesis may be taken as potentially the most hazardous to the embryo. Irradiation during yet *earlier* periods will presumably give all-or-none results: i.e., either very early death or normality of survivors. Irradiation at developmental stages *later* than the period of major organogenesis produces few gross malformations.

This statement appeared in a report to the United Nations Conference on the Peaceful Uses of Atomic Energy, in 1955.

The report goes on to point out that there are two possible common sources of radiation that may endanger pregnant women during peacetime: occupational exposure and medical exposure. Occupational exposure is not likely to be hazardous to a child before birth if it does not exceed the official permissible weekly dose. This would amount to less than 1 *r* during any critical period, and it seems unlikely that doses of less than 1 *r* would cause measurable damage.

In medical irradiation, which are X rays in doses large enough to constitute therapy, the doses are likely to be quite high, and they are therefore a definite hazard to a child before birth.

Dental X rays are another source of danger to the child. Dr. Martin L. Stone and his colleagues have recently pointed out that if pregnant women are to be urged to visit their dentists early in pregnancy, they should also be warned at the same time of the hazard of allowing the X rays to hit the pelvic area. To sit in a dentist's chair while fifteen or twenty X-ray exposures are being made, with the rays directed downward toward the pelvis, is to expose the reproductive organs and the developing child to large doses of irradiation.

Uninspected dental X-ray machines may deliver as much as 315 *r* at a single sitting. Dr. W. E. Nolan and H. W. Patterson investigated this matter in the San Francisco area and reported that some dental X-ray machines there delivered 113 *r* and 280 *r* at a single examination.

Radioactive substances are sometimes used as medicines in the treatment of various diseases. *Pregnant women should not use them.*

Radiophosphorus and radioactive iodine are two of these substances, and laboratory demonstrations have shown that both of these can be seriously damaging to a developing embryo. Radioactive iodine is perhaps the more widely used of the two, because of its value in the treatment of goiter. The danger of this treatment when a patient is pregnant is that the child's thyroid gland may be capable of absorbing a large dose of the radioactive iodine, with the possible result that the gland may be destroyed. This, of course, can cause numerous serious difficulties.

Some investigators, notably Dr. E. M. Chapman and his co-workers, have questioned the validity of this threat. They point out that a child's thyroid gland does not even begin to function at all until the beginning of the fourth month of gestation. And they report treatment with radioactive iodine in one woman in the second month and another in the sixth month of pregnancy with no apparent ill effects on their children.

Nevertheless, radioactive iodine is unquestionably harmful to a child before birth in other ways. Numerous studies and observations have demonstrated that cretinism and sterility have resulted from this treatment.

Pregnant women and their physicians are foolhardy to expose a child before birth at any stage whatever of his development to radioactive materials of any kind.

One of the most interesting reports in the entire field of radiation, published in 1959, is concerned with the radioactive elements in the rocks of New York State, and their possible effects on the rates of congenital malformation in that state. Drs. J. T. Gentry, E. Parkhurst, and G. V. Bulin, Jr., conducted the study and correlated the two factors.

Between 1948 and 1955, there were 16,369 children born in New York State who showed some kind of congenital malformation. This is an average annual incidence of 13.4 for each 1,000 live births. The rate for the United States as a whole is about one-half of this. The investigators asked themselves: Why?

Studying the statistics in greater detail, they discovered that there were significant differences within the state. For example, about one-fifth of the townships in the state had a malformation rate of twenty out of every 1,000 births, or higher, whereas other localities had much lower rates. Again, the question arose: Why?

The answer was found to lie in the deposits of highly radioactive rocks and soil distributed unevenly around the state. Using independent studies of geological data, the investigators classified all the townships in the state as to the probable presence or absence of rocks and soil with high levels of radioactivity. They labeled each township either "probable" or "unlikely," as a source of radioactivity. Then they correlated these labels with the rates of malformation in each township.

For all the areas classified as "probable," the incidence of malformation was 15.8 for each 1,000 live births. For the "unlikely" areas, the rate was 12.9. The most highly radioactive materials were found to be those areas with outcropping igneous rocks. These areas had a malformation rate of 17.5 per 1,000. Areas with extensive glacial deposits had a rate of 15.4 per 1,000.

In other words, the more radioactivity, the higher the number of children born with malformations.

Independent measurements were made by these investigators of the levels of radiation at various places about the state: the levels lay mainly between 8 and 12 microroentgens per hour, a microroentgen being 1/1,000,000 roentgen. Over a thirty-year period, this would amount to a dose for each individual in that area of something between 2.1 r and 3.2 r.

Is this enough to build up mutations over several generations? Apparently it is. Dr. Gentry and his co-workers investigated the length of time families had lived in the various areas of the state. In the areas with a high rate of malformation, some member of 20 percent of the families had lived in the area for over 75 years. In the areas with a low rate, some member of only 12 percent had lived there as long. This strongly suggests that mutations may have been taking place over several generations.

A note of caution is needed here: although there is no question that there is a connection between radioactive rocks and soil and the high rates of malformation in that state, there is as yet no proof that the one causes the other; it is proof only that the two are associated. Indeed, other studies have demonstrated this same kind of association in other parts of the United States. A preliminary study conducted for the U. S. Atomic Energy Commission by Drs. J. Kratchman and D. Grahn reported evidence that connects deaths from malformation with major deposits of uranium ore, uraniferous waters, or helium. This study is not definitive; the investigators themselves point out that many factors have not yet been studied in connection with this subject. Nevertheless, they find their conclusions provocative and worthy of further investigation.

One of the interesting aspects of studies of this kind is that large populations can be carefully scrutinized for the effects of very low levels of radiation exposure. Just as minor brain damage from a specific cause is difficult to measure in an individual child but can be studied in large numbers of children, so specific malformations—which may be caused by any one of several factors—are easier to pinpoint in large numbers than in small.

At best, studies of this subject are difficult to carry out. How does one differentiate, for example, between the effects caused by an increased mutation rate and those due to direct radiation? Also, which of these variables is the critical one: sex ratio, death rate of children before birth or after birth, birth weight, frequency of illness? Perhaps one of these is the principal consequence of radiation, rather than the rate of malformation. Many years and many dedicated investigators will be needed to work through the enormous body of information waiting to be studied.

Sterility may be another consequence of radiation. We are all familiar with the fear of sterilization of adults who were previously fertile; in fact, there is a case on record of a young man of twenty-four, who was rendered sterile by treatment with radioactive iodine. But what about producing sterility in children before birth? This, too, appears to be possible. Drs. L. B. and W. L. Russell have demonstrated this with female laboratory animals and have found that the effects of radiation vary considerably with the stage of development of the animal at the time she is exposed to the X ray. The reproductive cells of a mouse, for example, are most sensitive to radiation two weeks after birth; even a small dose of radiation at this stage will cause the animal to become sterile after producing a single litter, or perhaps a second smaller litter. At this age, the reproductive cells in the mouse have just reached the stage in which they will remain throughout most of the animal's adult life.

We cannot be certain that the human ovary is equally sensitive to radiation at a comparable stage of development, but scientists have found reason to believe that it is.

In the human female this stage is reached, not after birth, but at some period during gestation. Therefore, we must assume—at least until it has been proved false—that even quite low doses of radiation can produce sterility in human beings before birth. These doses may be much lower than we would have guessed from the effects of radiation on the cells of the adult.

In the laboratory animals, the consequences varied widely with the age of the animal. Exposure on the day of birth, for instance, appeared to have no effect on the later fertility of the animals. This was true even when the dose was large and given at a high rate. These animals commonly produced nine or ten litters when they reached maturity.

Adult animals, given the same low dose that produced sterility when it was given to the two-week-old animals, were also not affected by it. They later produced an average of 13.3 litters. But when adults were given acute doses, such as those given without any effect on the newborn animals, they produced no more than two litters and were afterward sterile.

Radiation also appears to be connected with the incidence of leukemia and other malignancies in children. A preliminary report by Dr. Alice Stewart and her co-workers published in 1958 showed that mothers who had been irradiated during pregnancy were twice as likely to have children who later developed infantile leukemia than those who had not. This study involved 547 children under ten years of age who died of leukemia. In this group there were several sets of twins in which only one of the pair was affected. Possibly the reason for this was the position of the irradiated twin, in front of the other, where he would absorb the greater part of the dose of radiation.

Other studies have followed. Many of them, involving many thousands of children, appear to show that malignant diseases in children are associated with radiation before birth, and many investigators have concluded that an apparently harmless X ray of a mother's abdomen while she is pregnant may, indeed, be the cause of later cancer or leukemia in her child.

On the other hand, additional studies have failed to verify such a connection. Some groups of children with leukemia have not been irradiated more frequently than children without it. The difficulty with these studies is that they have been based on samples too small from which to draw statistically significant conclusions.

One piece of evidence that is convincing to many people is the increase in the rate of leukemia in Britain. The death rate from leukemia there has risen from 1.1 to 5.3 in every 100,000 people in the last twenty-five years, and the most rapid rate of increase has occurred recently. In the four years preceding 1959, deaths from leukemia over the whole country increased by 13 percent. Dr. T. A. Phillips, who made this study, has concluded that a possible cause of most of the recent increase is strontium-90 fallout.

The death rate from leukemia varies from locality to locality in some parts of the world, and an attempt has been made in Scotland to relate this variation to background radiation. For example: The mean annual exposure to radiation of the population of Aberdeen is significantly greater than that of Edinburgh. Death from the two forms of leukemia which we know can be produced by radiation—acute leukemia and chronic myeloid leukemia—are more frequent in Aberdeen than in Edinburgh—approximately 300 as against 226.

Cases of leukemia and bone cancer during the next seventy years which may result from exposure to fallout from nuclear testing through 1961 have been estimated to be something under 2,700. This estimate was made in January, 1963, by Dr. C. L. Dunham, of the United States Atomic Energy Commission.

Another possible effect of radiation, not yet studied in man but clearly established in animals, is to shorten life. Even moderate doses of radiation shorten the life span of laboratory animals, and the reduction is apparently in proportion to the amount of radiation. Dr. J. B. Hursh and his colleagues and Dr. W. L. Russell found that 1 r of exposure leads to a shortening of the life of a rat by about one ten-thousandth; translated into the life of a man, this means that an exposure of 1 r would shorten a human life by about 2.5 days, or about one month for every 10 r. The present permissible weekly exposure of 0.3 r would lead to a shortening of about 7½ years.

Radioactive elements can also contaminate our food. Fallout of radioactive iodine, barium, strontium, and cesium penetrates the soil, passes into the grasses, from there into the bodies of the grazing cattle, and from there to the human being who eats the meat and drinks the milk. A pregnant woman can pass these along even further —to the body of her child. In fact, strontium can even be stored in her bones during a period when she is not pregnant, and some of it will still be available to damage a child who is conceived sometime later. The most dangerous time of exposure to fallout contamination, however, is during the last three months of pregnancy, when the bone formation of a child is most active.

In 1959, the director of the Laboratory of Radiation Biology at Cornell University, Dr. C. L. Comar, said that at that time the levels of radioactivity in human beings were not high enough to produce any observable effect. He added, however, that "indiscriminate testing of nuclear weapons could lead to levels of strontium-90 in the food chain that would be of definite concern." Three years later, in 1962, levels of radioactivity had more than doubled, and in at least one American town, Palmer, Alaska, the iodine-131 in milk rose above the safe level for infants and young children.

The atomic bomb explosions over Hiroshima and Nagasaki in 1945 exposed many pregnant women and their children to varying amounts of radiation; those who were less than a mile from the center of the blast in Hiroshima received perhaps 350 r. Dr. G. Plummer has reported that seven out of 205 of these children, or 3.4 percent, were seriously affected. In Nagasaki the results were more severe: among the children of thirty women who were within 3,000 meters, or slightly under two miles, of the center of the blast, there were thirteen deaths, either before, during, or shortly after birth; this is 43.3 per cent. Among the seventeen surviving children there were four who were mentally deficient. The mothers suffered severely from this radiation, too: the rate of illness and death among them approached 60 percent. This is in sharp contrast to a group of mothers who were within 2,000 meters of the center of the explosion but who did not show signs of major radiation; the rate of illness and death among them was 10 percent.

The Atomic Bomb Casualty Commission reported on all the children they could find and study who were exposed to the explosions. Of 325 children whose mothers were pregnant at the time and who were within a radius of 2,500 meters, nineteen children were retarded, mentally and physically. This is 6 percent. Most of these nineteen were in the third or fourth month of their prenatal life when they were exposed, and at a distance from the explosion of between 725 and 1,624 meters. Sixteen of the nineteen mothers had had symptoms of radiation disease.

Another effect of radiation in large doses over short periods of time was clearly demonstrated by these atomic bomb attacks: the serious weakening of active immunity to disease. The normal balance of microflora in the body is disturbed, and both red and white blood cells are affected; this adds up to a reduced resistance to disease. On the other hand, prolonged low-level radiation appears not to affect immunity at all.

All the studies on radiation, from atomic bombs and X rays alike, underline the dangers to a child before birth of exposure to radiation. The amount and kind of damage to the child depends on the amount of exposure and the period of his development when he is exposed, but even a small amount can have serious consequences.

The lesson for a pregnant woman in these studies is a simple one: radiation is dangerous to a child before birth; his mother should avoid it. If she must be X rayed for diagnostic purposes, her child is safe only if the rays are directed away from her pelvic region. This part of her body should never be exposed to X rays, even the smallest amount.

In fact, even a woman who is not pregnant but who is still in her childbearing years should never allow therapeutic irradiation of her ovaries unless a severe disorder is involved and she cannot be treated in any other way, and diagnostic X rays of the pelvis should be kept to a minimum.

No individual woman can control the amount of radiation she is exposed to from fallout or from the naturally radioactive materials in her surroundings, but she has some control over X rays. She must exercise that control on behalf of her children.

XV—THE BRIDGE: BIRTH

THE MORE I observe, the more I realize that none of the developments and changes which life brings find the individual so well prepared „as for birth." This statement was made by Dr. Sandor Ferenczi, the distinguished Hungarian psychoanalyst.

And yet birth is potentially dangerous. Indeed, the most dangerous journey in life is through the four inches of the birth canal, as Dr. T. N. A. Jeffcoate, Professor of Obstetrics at the University of Liverpool, has put it. Each of these statements sums up a complex series of events in a simple sentence. And like most such concentrations of meaning, each is an oversimplification, and to some extent each is misleading.

A prospective mother, looking forward eagerly to the end of her pregnancy and searching for an understanding of the birth process, might begin by combining these two statements in her mind: birth is indeed hazardous; and almost all children are well prepared for it.

Birth is not a beginning: we have seen again and again in the course of this book that the true beginning is at conception. Nor is birth an ending. It is more nearly a bridge between two stages of life, and although the bridge is not a long one, a child crosses it slowly, so that his body may be ready when he steps off at the far end.

Birth is a long process. An average first labor consumes fourteen hours; the average for later labors is eight hours. Does this seem overlong for a journey of only four inches? Perhaps; still, when we gain some understanding of the operations that are going forward during those hours, the time span seems appropriate enough. No satellite launching, with its highly trained personnel and its electronic devices, was ever so beautifully timed as a normal, ordinary, uneventful human birth.

In general, the various physical changes that precede and accompany birth accomplish a single specific purpose: they prepare the child to function as an independent individual in several areas where until now he has been dependent on his mother. During his trip across the bridge, the child's various sustaining systems are being stimulated to begin their operations. By the end of the trip, the systems are ready to begin functioning, the child is ejected, he gives a cry and begins to breathe: he is born.

One of the first happenings of importance in the series of events is that the oxygen level in the placenta drops precipitously. This leads to two significant consequences: the first is that the placenta ceases manufacturing the hormone progesterone. This hormone has been present during pregnancy, and its function has been to maintain the pregnancy. In fact, its name comes from two Greek words meaning "in favor of" and "gestation." And this is the hormone that is sometimes given artificially to women who are threatening miscarriages. When the amount of progesterone in the system diminishes, the muscles in the uterus respond by beginning the contractions that are known as labor. Thus the birth process begins.

The drop in the oxygen level of the placenta accomplishes another purpose, too. It decreases the efficiency of the placenta as a lung for the child. Since the fourth month of pregnancy, the placenta has been acting as the child's lung: it has transferred the oxygen from the mother's bloodstream to the bloodstream of the child, and it has transferred the waste products from the child's circulation to the mother's. When it suddenly becomes unable to perform this function adequately, the child's own system is alerted, and a series of complex procedures is set into motion that will end with independent breathing and circulation in the body of the child.

An opening in his heart, for example, begins to close, and an arterial canal begins to shut down. These two actions force the child's blood into a new path, and instead of traveling directly from one side to the other of the heart, it now begins to use the arteries and veins between the heart and the lungs. In order to use his lungs, however, the child must be breathing air; also involved are the functioning of the muscles of his chest and abdomen, his diaphragm, and the whole of his upper respiratory system. Some of this must wait until other parts of the body are ready.

In the meantime, the other systems of the body are preparing to function independently: the digestive system, which takes in food and liquid and breaks it down so that it can be carried through the body by the bloodstream and absorbed by the body's tissues; the eliminative systems, which carry the waste products from the intestines, from the blood by way of the kidneys and the urinary tract, and from the skin through the sweat glands; the nervous system, which enables an individual to make the proper responses to the things that happen to him; and the endocrine glandular system, which plays an important role in growth, development, and behavior, and also helps in the functioning of all the other systems.

While each of these systems is getting ready, the uterus continues its rhythmic contractions. The child is being propelled slowly down the birth canal. The amniotic sac, known as "the bag of waters," breaks, and the amniotic fluid is expelled. Now, for the first time in his life, the child is exposed to air.

Changes in pressure and temperature take place immediately. The diaphragm begins a bellowslike action. This helps to inflate the lungs with air, and they press upon each side of the heart. Sometimes these changes happen while the child is being born; sometimes they occur after he has been delivered.

In the meantime, the child is being assaulted on all sides by the contracting uterus. His head is progressively pressed and squeezed against the walls of his mother's bony outlet. In addition, he may be having a hard time breathing oxygen. Sometimes, at this stage, a child can be heard to cry, even before he is actually born. It is possible that this is a cry of pain, but it is more likely simply the sound that accompanies the expulsion of air from the child's lungs, the cry of the newborn. The child's oxygen level is low, and this means that his perception of pain is also low; although his situation is fraught with uncertainty and discomfort, he is probably not conscious of pain.

With his cries immediately after he is born, the child

signals that he has survived the most difficult stage of his journey. At this point, his mother should be allowed to see him, and to nurse him, and to assure herself that he is healthy and normal. She will then be able to relax and rest after a good job of work.

The placenta will be ejected within a few minutes, the bleeding from the torn vessels of the uterus will begin to decrease, and the uterus will contract. The child, by now, will be breathing by himself, his various other sustaining systems will be functioning, and he will be ready to participate actively in the next stage of his development.

We have all been through the process of birth in the first person. Most of us have participated in it in other ways: as mother, father, doctor, nurse, investigator, observer. From experience, from reading, from observation, we know the outlines of what happens during birth. And by stretching our imaginations, we think we understand some of the aspects of the process.

Our scientists certainly know how important it is to an individual's physical condition to have a smooth birth. Again and again in this book, we have seen evidence that a child who has had a good prenatal life, and is headed for a good postnatal life, can be seriously harmed by some accident during the process of birth. His oxygen supply is perhaps the most crucial single influence at this time, and a prospective mother approaching the time of her child's birth may benefit from a rereading of the section of this book devoted to oxygen. All of Chapter IV, but particularly pages 49–52, is especially pertinent.

We know, too, that some other influences make their presence known indirectly during the process of birth. Malnutrition is surely the most important of these. A woman who has been poorly nourished throughout her pregnancy—perhaps throughout her life—is likely to be poorly developed in all her organs and functions. The complex series of actions and reactions that makes up her part of the birth may be performed inadequately. Furthermore, her child, whose nutrition has been dependent on her and whose organs and functions are also poorly developed, may be less able to do his part well. Thus the various influences that may have been quiescent during the pregnancy come together at the crisis of birth. We know that they are important, and we understand the ways in which they act.

Some aspects of the process of birth are more complex, and even surprising. It has occasionally happened that something we thought was simple and clear has turned out to be not so at all. Consider the function of the contractions of the uterus. This appears to be straightforward and obvious: the uterus contracts steadily and rhythmically as it does in order to propel the child down the birth canal. But is this all?

These contractions also stimulate the child's skin; is this important? I have elsewhere presented the evidence indicating that it may be extremely important, particularly in activating some of the systems of the child's body. Particular mention may be made of the gastrointestinal system, the genitourinary system, and the respiratory system. I have suggested that without the stimulation of the skin given by the contractions of the uterus these systems might not function well.

How does one test such an hypothesis? First, we consider the other mammals with which we are most familiar: is there stimulation of the skin of the newborn anywhere among them? Immediately, of course, we find it in the licking that the mother of a new litter of animals gives to her young. She spends a great deal of time at this licking. Furthermore, much of her attention is devoted to the ano-genital region of the young animals. Looking at this activity from the point of view of human beings, we long ago decided that the mother animal was washing her young, that she was primarily concerned with cleaning them. This was an obvious explanation and a satisfactory one.

But what happens if animals are not licked? Do they simply remain dirty, or is there some other effect?

The answers to these questions have been known for years to animal breeders, veterinarians, and the staffs of zoological gardens: a newborn mammal that is not licked is likely to die because it is incapable of bowel and bladder elimination. In other words, two important systems of its body, the gastrointestinal and the genitourinary, fail to function; as a consequence of this, it dies. Although this knowledge has been known to the men who work with animals, for some reason it has not found its way into the medical forum.

The mechanism of this is not so farfetched as it first appears. The peripheral sensory nerves in the skin are stimulated by the licking, and in human beings by the contractions of the uterus lasting over many hours. These nerves conduct the impulses to the central nervous system. There, at the proper levels, they are shunted out again to the various organs of the body. When the stimulation of the skin has been inadequate, the nervous system is inadequately stimulated and the principal organs fail to act.

The most interesting evidence to back up this hypothesis originated in a laboratory, where scientists were trying to raise germ-free animals. Professor James A. Reyniers was in charge of the Laboratories of Bacteriology at the University of Notre Dame, where he and his colleagues encountered great difficulty in the early days of their work, because their animals died. Apparently the deaths were caused by the failure of the animals' gastrointestinal and genitourinary systems. The investigators were completely defeated by this problem until one of the workers on the project, a former member of the staff of the San Diego Zoological Gardens, solved it. Remembering the kind of care that newborn mammals received from their mothers, she suggested that the investigators at Notre Dame stroke the perineum—the region between the anus and the external genitals—of each newborn animal after each feeding until it defecated or urinated.

Dr. Reyniers later wrote in a personal communication:

Rats, mice, rabbits, and those mammals depending upon the mother for sustenance in the early days of life apparently have to be taught to defecate and urinate. In the early period of this work, we did not know this and consequently lost our animals. The unstimulated, handfed, young died with an occlusion of the ureter and distended bladder. Although we had for years seen mothers licking their young about the genitals, I thought that this was largely a matter of cleanliness. On closer observation, however, it appeared that during such stimulation, the young defecated and urinated. Consequently, we started to stroke the genitals of the young after each hourly feeding with a wisp of cotton and were able to elicit elimination. From this point on we have had no trouble with this problem.

When a newborn child fails to breathe, an age-old practice is to slap him smartly once or twice on the buttocks; this will often stimulate him to breathe. Another practice is to put him into alternating hot and cold baths. Both of these methods stimulate the skin, and this in turn activates the respiratory system.

Evidence to support this hypothesis also comes from groups of premature children. These children are frequently born after very short periods of labor; when this is the case, the child has had much less stimulation of his skin from the contractions of the uterus than the full-term child, who is born after a labor of eight hours or more. Independent studies have shown again and again that premature children have much more difficulty with their breathing than children born at term. Also, they achieve bowel and sphincter control later and with much greater difficulty. And they have more frequent and greater feeding problems. The indications are that inadequate stimulation of these systems during labor results in inadequate development of their functions.

Another common problem of premature children, known as the meconium plug syndrome, points in this same direction. This is the condition in which the intestines of the newborn child are not emptied as they normally are, and the material forms a plug and stops the passage of food through the intestines.

These difficulties of premature children may well be caused, at least to some degree, by the fact that their skin has not been sufficiently stimulated during the process of birth. Further investigations will have to be conducted, of course, before this can be settled one way or the other, but the hypothesis itself opens the door to other interesting possibilities. We may find, over the next few years, that many things we thought we had understood—such as the licking of newborn animals by their mothers—we are actually far from understanding.

Something of the importance of the normal birth process as a preparation for postnatal function is indicated by what happens in the case of Cesarean-delivered babies who have not had the advantage of delivery through the birth canal. In normal birth the chest of the baby is compressed as it passes through the birth canal. This induces considerable pressure within the chest cavity, the heart, circulatory system, lungs, and diaphragm. Furthermore, by its contractions the uterus squeezes the blood from the placenta through the umbilical cord and so into the baby. In the absence of such prolonged contractions, the baby will not receive as much blood. The consequences of all this will be that there will be a failure to expel amniotic fluid from the baby's upper passages, and there will be inadequate aeration of the respiratory tract as well as inadequate stimulation of the circulatory system.

Other aspects of the birth process we know we do not understand, and some of them will probably elude us for a long time. Some, in fact, will demand that we bring to them more perception and intuition and intellectual courage than most of us are willing or able to muster. One such aspect is the psychological significance of birth.

Is the process of being born such a shocking experience to human beings that we carry the effects of it within us all the rest of our lives? Many psychologists believe that it is.

Freud calls birth "such a concatenation of painful feelings, of discharges of excitation, and of bodily sensation, as to have become a prototype for all occasions on which life is endangered, ever after to be reproduced again in us as the dread or 'anxiety' condition." In other words, birth threatens life, and whenever in later years an individual is in danger—whether it is real or imagined—he relives the period of greatest danger of all. The interruption of his breathing was the cause of his anxiety at birth, suggests Freud, and always after that when he becomes anxious his breathing tightens, although there is no physical reason for it to do so as there was at birth. He also believes that it is suggestive that this first state of anxiety in a human being occurs when he is first separated from his mother.

Freud adds an interesting personal note to this suggestion in his *Introductory Lectures on Psycho-Analysis*.

It may perhaps interest you to know how it was possible to arrive at such an idea as this—that birth is the source and prototype of the anxiety effect. Speculation had least of all to do with it; on the contrary, I borrowed a thought from the naïve intuitive mind of the people. Many years ago a number of young house-physicians, including myself, were sitting around a dinner table and one of the assistants at the obstetrical clinic was telling us all the funny stories of the last midwives' examination. One of the candidates was asked what it meant when the meconium (the child's first excreta) was present in the waters at birth, and promptly replied: "That the child is frightened." She was ridiculed and failed. But I silently took her part and began to suspect that the poor unsophisticated woman's unerring perception had revealed a very important connection.

Birth, according to Freud, produces a sense of helplessness in the individual. Through this sense of helplessness he becomes aware of the dangers of the world, and the importance to him personally of these dangers. Also, through this sense of helplessness he develops dependence. Thus is born the need for love, which, as he says, "will nevermore leave the individual."

Otto Rank developed Freud's theory of anxiety in a series of books beginning with *The Trauma of Birth,* published in 1924. Rank sees the life before birth as a blissful state of security and satisfaction. Birth is a rude interruption of that bliss, a destruction of all security. This is a tremendous psychic shock, or trauma, from which the individual spends his life trying to recover. On the deepest levels of his being, in fact, he never accepts the severance from the warm, nourishing, protecting mother. The whole of his later life can be viewed as reaction to the loneliness and suffering of existence outside the womb. According to Rank's beliefs, a human being's entire adjustment to his environment consists in his attempt to overcome the trauma of birth. Some individuals succeed better than others; neurosis is the name we give to the attempts that have failed.

Other psychologists have looked at other aspects of birth. Dr. Phyllis Greenacre points out that during birth a child is the center of a great number of events of various kinds and intensities. His reactions to most of these are primarily muscular, but because of his cramped position and his inability to move more than a small amount, he cannot respond to these events as actively as he would like to do. Dr. Greenacre suggests that this builds up a considerable amount of tension in the child, and this "may conceivably be incorporated into the working balance of the individual and become temporarily or permanently a characteristic of his make-up." She suggests that the tension that is built up in a child during birth may determine to some degree the drive and the energy of that person in later life.

Another scientist studies the dreams of his patients and finds there evidence that life before birth is a state of "uterine bliss" and that birth itself is traumatic. Dr. Nandor Fodor hypothecates that "the changeover from pre-natal to post-natal life involves an ordeal as severe as dying. Hence, the fear of death begins at birth and is based on a maelstrom of bewildering experiences that are covered by infantile amnesia but break through in nightmares or become converted into symptoms."

Other psychologists have cited a recurring childhood dream as evidence that the experience of birth is carried in the subconscious well into postnatal life. This is the dream that takes the form of finding oneself in a dark narrow chimney, struggling to descend. This is often accompanied by terrifying feelings of suffocation.

Not everyone, of course, agrees that birth is a shocking experience. The statement with which this chapter opens, by Dr. Sandor Ferenczi, is a strong expression of the opposing view. Indeed, any observer must concede that from the physical point of view, at least, a human being is well prepared for birth when it comes.

Others have pointed out that no real evidence has ever been produced for believing that the situation of a child before birth is a supremely pleasurable one. In addition, others object that "uterine bliss" can have no conscious meaning for the child, since he cannot use the higher centers of his brain in any case until three months after birth. Furthermore, since during the birth process the amount of oxygen reaching his brain is much reduced, it is unlikely that the experience would even be recorded in it.

These comments and observations are provocative. However, they are not so much arguments for or against specific beliefs as they are almost random shots at an enormous, unseen target. The target is so large, in fact, and so distant, that we cannot even be sure that we are aiming at the center of it. We are a long way from a true understanding of the psychological significance of birth.

Investigators in laboratories and hospitals are approaching this subject from another angle, by collecting evidence from delivery rooms and nurseries. These scientists are beginning to show us that the process of birth contributes to an individual's emotional condition after birth, to an extent that can be measured.

One method of doing this has been to compare children who have been delivered by Cesarean section with those who have been born through the birth canal. Dr. Marion Kenworthy has reported that the Cesarean-sectioned child is prone to be less sensitized than the firstborn child in a normal delivery. "He cries less," she says, "is markedly less irritated by the contacts of handling."

On the other hand, a premature child appears to be more nervous and more irritable than a full-term child. His birth, according to Dr. Mary Shirley, is likely to have been "cataclysmic, unduly prolonged, or precipitant," and the implication is that this has resulted in his nervous irritability.

Several investigations have been made among children who had been delivered with instruments. Dr. P. L. Schroeder found that their most characteristic personality traits were distractibility and hyperactivity. In a group of thirty-five children of nursery school age, all of whom could be called "anxious," Dr. Louise Despard found that nineteen, or 54 percent, had been delivered with instruments. The presumption here is that children delivered with instruments have a more difficult transition from prenatal to postnatal life than other children, and this is shown in their personalities and temperaments. Stuttering, too, has been connected with instrument birth. Dr. J. L. Boland studied a group of 209 stutterers and found among them a greater number of children who had been born with the aid of instruments than is found in the general population.

Does this mean that the use of instruments during the birth of a child will cause a baby to be emotionally less healthy than he would have been if the instruments had not been used? No. In fact, the work that has been done so far in this field means nothing, yet. Assigning causes and effects in such a complicated affair as the birth of a child is much too difficult to do so easily. A premature child, for example, may indeed be more nervous and more irritable than a full-term child, as Dr. Shirley reports. But the nervousness is not necessarily caused by the prematurity. They may both be caused by some third factor. Or the prematurity may contribute to a state of nervous irritability that would not exist at all if it were not for some other influence altogether. We cannot say yet that we understand any of these things.

Still another aspect of birth that we are only beginning to understand concerns an extension of the period of gestation. Customarily, we think of this period as ending at the birth of the child. Literally, this is the proper interpretation. We are coming to see more and more clearly, however, that a newborn child is almost as dependent on his mother as he was before he was born. For a length of time approximately as long as his prenatal period, he will continue to be truly helpless, unable to protect himself, unable even to move away from an approaching danger. When he is finally able to crawl, sometimes as early as five months, he has arrived at a stage that many other animals achieve before birth. Therefore, it is not unreasonable to look upon these first months of his life as a continuation of the period of gestation, and his birth as the bridge between two equal periods of rapid development and dependence.

We now begin to see the life of an individual as a continuum beginning with conception and ending in old age. At no point in that long span of years does a human being change suddenly from one stage to another. From babyhood to childhood to adolescence to youth to middle age to old age: it is all gradual. And so it is with the transition between the prenatal period and the first months of postnatal life. Although the human infant has already spent, on the average, 266½ days growing and developing, and has indeed succeeded in putting together a marvelous machine for living, he is still a pitifully immature creature when he is born. Functionally, he is undeveloped. Compared to the newborn of other mammals, he is dangerously dependent.

The length of gestation in every animal is a characteristic that has been determined by natural selection over many thousands of years, in response to the necessities of the life it leads. The lion, for example, can protect its young; the baby lions can therefore be small; the gestation period can therefore be relatively short—only 105 days. Professor John Bostock, of the University of Queensland, has discussed these matters in an extremely interesting study. He goes on to point out that cold-water mammals such as seals must be able to withstand cold water from the day of their birth, and by the time they are six weeks old, they must have learned to swim and be able to flee from danger; as a result, they must have a long gestation period, approximately 350 days. A baby elephant must be able to move with the herd immediately after birth; this demands a particularly long period of gestation, from 515 to 617 days. Deer must be able to run shortly after birth, and in order to do this they too require a long gestation period: 230 days. Man's gestation period is long, too, but in his case special conditions apply: his period before birth is determined by the size of his brain.

By the time a child is born, his brain has increased to such a size that his head can barely be accommodated in his mother's body. If it were any larger, the child could not be born at all; the size of a baby's head is already a hazard in some births.

And yet, his brain is not nearly as large as it will soon become. When he is born, the volume of his brain is about 350 cubic centimeters, or approximately 21 cubic inches. By his first birthday, it will have reached more than 800 cubic centimeters, or more than twice the volume at birth. Thus, if the child were to wait for his birth until he is as mature as a newborn fawn, for example, or a baby elephant, he would never be born at all.

The result of this situation is that man has worked out, over many thousands of years, a compromise: he is born with the maximum-sized head that can be accommodated in the birth canal, and a dependence on his mother that is greater than that of any other animal.

We are therefore justified in thinking of the first months of a child's life as a continuation of the period of gestation, and we are justified in treating the child and his mother, and the relationship between the two, as though they were still closely connected.

This concept opens the door to a new view of the nature of early infancy, and a mother who takes this view of her child and of her own role in the child's life may be prevented from making some of the more common mistakes with her newborn child. Admittedly, it is useful to divide life into little packets of time, and a woman who has just finished nine months of pregnancy may be excused if she concentrates on the pleasures of not being pregnant. But she should not allow these to blind her to the important fact that her newborn child is still almost as dependent on her as he has been since his conception, and he will continue to be so for some time to come. Like the little marsupials, the opossum and the kangaroo, which find their way into their mother's pouch and stay there until they are old enough to cope with life themselves, the human infant is born before he is ready. Unlike the marsupial, however, he has no warm pouch next to his mother's skin to lie in as he matures. He is put into a bed of his own, even a room of his own, and given a bottle to drink from, and left alone.

But he still needs his mother. He needs to be held in her arms closely and often, and spoken to softly, and nursed at her breast. He needs to hear, still, the reassuring sound of her heartbeat. He is a world away from her womb, but the physical distance is only a few inches, and the psychological distance is less than that.

His mother needs him, too. All the time she has been carrying him, her body has been gradually and elaborately preparing itself for the role that it is now required to play. This is to continue, on another level, the close relationship with her baby that has existed between them for nine months. In our superior wisdom, we in the Western world have come to disregard this preparation and to fail to take into account the importance of the continued relationship between mother and child.

Even after birth, whatever benefits one benefits the other. For example, when the child begins to nurse, he stimulates the flow of colostrum. This is the liquid in her breasts that comes before the milk, and its function for him is not only nutritional, but also to clean out the meconium from his intestinal tract, act as a laxative, prevent diarrhea, and, in addition to those obtained through the mother, to confer passive immunity against a number of diseases. In the 1920s Dr. Theobald Smith of New York University and his colleagues demonstrated that colostrum conferred immunity upon calves against colon bacillus, and in 1934 Dr. J. A. Toomey showed that similar immunization factors were present in human colostrum, as well as immunizing factors against other gastrointestinal bacteria. Since then numerous studies have served to show how important colostrum is for the newborn. What better reassurance for the infant can there be than to nurse at his mother's breast, what better promise of good things to come? For his mother, the sucking stimulates reflexes that produce massive contractions in her uterus and speed up all the other activities of her body that will take it back to normal.

Breast feeding is probably the single most important agency through which the relationship between a child and his mother can be kept at its best. Breast feeding is the next logical step after placenta feeding; it is a natural process, and to ignore it, as so many women in the past generation have done, is to do violence to the best development of the child.

There is no question that breast-fed babies are at an advantage over bottle-fed babies. They are significantly healthier, because colostrum and breast milk contain many general factors for immunity to disease. This seems to apply particularly to intestinal disorders and respiratory diseases. Mortality rates are lower in breast-fed children. In addition, growth is more harmonious, the development of the face, the palate, and the teeth is better, and even intelligence has been found to be higher in breast-fed children than in others.

Breast feeding is becoming more common with the present generation of young mothers than it has been for many years. This development is largely due to the efforts of an organization, known as La Leche League, made up of women who believe firmly in the value of breast feeding and want to encourage other women to practice it. Many women have been discouraged in the past by physicians and hospital nurses, and for that reason and others have encountered difficulties in trying to establish sufficient milk for their babies, but the presence of an organization of other women of similar beliefs has helped them to persevere and succeed. Prospective mothers who are interested in joining this organization can write for information to La Leche League International, Inc., 3332 Rose Street, Franklin Park, Illinois.

There are several other organizations for helping prospective mothers and fathers, and those who are already parents. The Childbirth Education League and the Childhood Education League are two of these. Many communities have maternity centers, some public and some private, but all devoted to the task of helping prospective mothers with their problems. These organizations sometimes hold classes, for fathers as well as mothers, and they try to help anyone who asks for information and guidance. Any woman who is going to have a child can benefit from some contact with one of these organizations.

When a child comes to the end of his prenatal life, and crosses the bridge, and is born, he has successfully completed the most dangerous journey in his life. The anxiety of his parents, natural enough while they waited, has given way to joy, and the discomfort of his mother during the last weeks of the pregnancy has melted into relaxation and rest. If the child has been cared for during these last nine months with love and common sense, he will be fully prepared to continue to grow and develop.

XVI—CONCLUSIONS

THIS BOOK HAS BEEN a progress report on our growing understanding of the ways in which a child can be influenced before birth. It is by its nature incomplete. Obviously, no single book of this size could cover all the burgeoning activities in our laboratories and hospitals, and even if such a book were attempted, it would be out of date by the time it reached the reader. On the other hand, this enormous volume of work has immediate importance. Children are being conceived at this moment, and these children are not going to wait for a final conclusion from a laboratory to begin their development. And yet their development might be greatly helped if some of the information in this book were made available to their mothers. It is for the mothers of these children, as well as the mothers of all the children to come, that this progress report has been written, and it is to them that it is principally addressed.

Three facts stand out clearly above the mass of hypotheses, figures, conjectures, conclusions, and recommendations:

Things happen to children before they are born.
Most of these things are good; some can be harmful.
A prospective mother can determine to a large extent what will happen to her child.

We have much to learn about all this. Again and again, in the course of reporting studies and investigations, it has been necessary to say "This work is not yet finished," or "This conclusion is still tentative," or "It is not clear whether *this* was caused by *that* or whether they were both caused by something else."

A prospective mother, however, does not need to wait until all the evidence is in to act on the evidence that is already clear. She cannot doubt, for example, that nutrition can mean the difference between good health and poor health, or even life and death, to her child. Nor can she be in doubt as to exactly what a good diet is. This is one of the few certainties in life. If she does not already know, from school or from her reading, her physician can tell her. The most important item, therefore, for a prospective mother is a good diet.

Next: oxygen. If she lives a sensible, healthy life, her child will almost certainly receive from her enough oxygen to supply the needs of his growing body. Specifically, a pregnant woman should avoid those common items in our environment which rob the blood of oxygen. This means:

No general anesthetics
No smoking
No drugs
No clamping or tying of the umbilical cord
No poisons, if it is at all possible to avoid them, of
    the kind found in our polluted atmosphere

There are also some general precautions for every pregnant woman. She should avoid:

Infections
Unnecessary emotional upsets
Obesity
Undue fatigue
X rays and radioactive substances of all kinds

A very young woman, or one who is over thirty-five, or a woman who has some functional disorder, should take particular care of herself during pregnancy. She should begin this care by discussing her situation thoroughly with her physician and preparing to follow his recommendations.

A good obstetrician will probably be aware of much of the work reported in this book, or at least of the general trends that this work has established. However, all this knowledge is recent, and some of it may not yet have become systematized and incorporated into the working capital of all competent physicians. A forward-looking woman may be able to help the cause of the next generation by discussing these matters with her doctor and spreading her information as widely as possible.

To take care of a child before birth is not a difficult task. He is so eager himself to grow and develop to the best of his capacity that the task of the adults in his life is primarily one of removing obstacles from his path. Now that we know what many of these obstacles are, a prospective mother need only live a healthful, relaxed life during pregnancy to insure the best possible chance for her child to be born normal and healthy, with the promise of a bright future.

DIET DURING PREGNANCY

A proper diet for a pregnant woman may be summarized as follows:

Milk: One quart daily, or a lesser amount if cheese is included in the diet. This should be American (Cheddar) cheese rather than cottage cheese, because the former has a much higher calcium content.

Leafy green and yellow vegetables: One or more servings daily. These should be very fresh or canned or frozen, and should be eaten either raw or cooked for as short a time as possible. Protracted cooking destroys vitamins.

Citrus fruits: Tomatoes and other foods rich in vitamin C, such as cabbage, salad greens, green peppers, and turnips.

Meat, poultry, fish, and eggs: One or more servings daily. One or two eggs should be included every day, liver or oysters once a week, and beef as often as possible.

Breads and cereals: Four or more servings a day. These should be whole wheat, dark rye, or enriched.

Potatoes, and other vegetables and fruits: Two or more servings every day.

Butter and salt: In moderation.

GLOSSARY

Abortion. The giving birth to a fetus before the twenty-eighth week of gestation.

Achondroplasia. Abnormality of cartilage conversion into bone arising in fetal life. Characterized by short arms and legs, relatively long body, good intelligence, big head and face, square nose with depressed bridge, and spadelike hands.

ACTH. The letters for adrenocorticotrophic hormone. This hormone is secreted by the pituitary gland, situated at the base of the brain. The main target organs of ACTH are the adrenal glands, situated one on the top of each kidney. ACTH acts on the outer "rind" of the adrenal glands, the cortex, to stimulate the addition of cortical hormones, such as cortisone, to the circulation. In this manner the organism is enabled to deal with stressful conditions in the external environment, because the metabolic changes thus induced provide the necessary energizing.

Allergy. Any abnormal reaction in the form of a hypersensitivity to any exciting substance or antigen.

Amniocentesis, transabdominal. Removal of amniotic fluid from the amniotic sac by needle syringe through the abdominal wall.

Amniotic sac. The bag of waters, the inner sac of the amnion, enclosing the fetus in the uterus. The amniotic fluid is contained within the sac, and provides the fluid medium in which the fetus lives during its gestation. The amniotic fluid is refreshed about every two hours.

Analgesic. A drug producing loss of sensibility to pain.

Ano-genital region. The region including the anus and the external genitalia.

Anoxia. Oxygen lack; the state of having less than the physiologically normal amount of oxygen in organs or tissues.

Antenatal. Occurring or existing before birth. Often used as equivalent to "prenatal," but in fact not quite the same. "Antenatal" refers to the whole condition of pregnancy; "prenatal" refers specifically to that condition as affecting the fetus.

Antibody. Any of various substances produced by the body as a reaction to contact with an antigen, and having the specific capacity of neutralizing or reacting with the antigen.

Antigen. Any foreign material which when introduced into the body causes the production of antibodies.

Anus. The opening at the lower end of the digestive tract for the elimination of solid wastes, situated between the buttocks. Often erroneously called "rectum," a term properly applied to the last part of the large intestine, which ends at the anus.

Autonomic nervous system. That part of the nervous system not usually under conscious control which is the active (motor) nerve supply of the smooth muscles, such as those of the digestive tract, the blood vessels, etc., and the glands.

Bacterium. One-celled organism belonging to the plant kingdom. In shape it may be spherical, rodlike, or spirally twisted. Most bacteria are harmless, but some cause diseases, including such serious diseases as tuberculosis, diphtheria, typhoid, and pneumonia.

Cardiovascular. Relating to the heart and blood vessels or the circulation.

Cell. The living active unit of all plants and animals, consisting of many specialized parts. In the nucleus of the cell lie the chromosomes.

Cesarean section. The operation by which a child is removed from the uterus through an incision in the abdominal and uterine walls.

Chromosomes. One of a number of thread-shaped bodies situated in the nucleus of plant and animal cells and carrying the hereditary particles, the genes.

Congenital. Existing at birth. To be distinguished from "hereditary," which always refers to a trait that has been genetically determined. A congenital trait may or may not be of hereditary origin.

Cortisone. A hormone secreted by the adrenal cortex, under the stimulation of ACTH, enabling the organism to respond to stressful environmental conditions, and involved in the regulation of salt and carbohydrate metabolism and water balance.

Cretinism. A condition due to lack of the thyroid gland secretion, thyroxin, beginning in fetal life, and becoming manifest at birth or in earliest infancy. Without treatment there is a marked arrest in bodily and mental growth, the subjects of the disorder being dwarfed, apathetic idiots. Early treatment with thyroid extract will usually result in the attainment of an approximately normal child and adult.

Cyanosis. The dark bluish or purplish color of the skin and mucous membranes due to deficient oxygenation of the blood in the lungs or to an abnormally great reduction in the volume of blood passing through the smallest blood vessels, the capillaries.

Development. Increase in complexity. To be distinguished from "growth," which is increase in size. The two usually go together.

Diabetes mellitus. A disorder caused by the failure of the pancreas to secrete adequate amounts of insulin. The tissues are unable to utilize sugar, and abnormal amounts appear in the blood and are excreted in the urine. There is intense thirst and water intake, with excessive output of urine. There is also increased susceptibility to infection. Wounds and injuries heal very slowly; there is a tendency to develop skin ulcers; circulation is impaired, especially in the extremities; and dental and visual defects are common. Death may result from coma due to the disturbance in fat metabolism and the excess carbon dioxide in the blood, resulting in air hunger.

Differentiation (of cells). The process whereby cells of a uniform type to begin with, develop varied and specialized functions differing from the original and other cell types.

Disease. An acquired morbid change in any tissue or tissues, caused by a specific microorganism and producing characteristic symptoms. To be distinguished from "disorder."

Disorder. A disturbance of structure or function or both due to a genetic or embryological failure in development or as the result of external factors, such as certain chemical substances, injury, or disease. May be inborn or acquired.

Down's syndrome. This was formerly called "Mongolism" or "Mongolian idiocy," an objectionable and misleading term based on a fancied resemblance to the Mongoloid peoples. The term "Down's syndrome," after Langdon Down, who first described the condition, is now preferred. Down's syndrome is characterized by severe mental and physical retardation, the presence of inner eyelid folds, a large space between the first and second digits of the hands and feet, stubby hands and fingers, a large tongue, etc. Chromosomal anomalies are usually associated, especially of chromosome No. 21.

Ductless glands. *See* Endocrine glands.

Edema. Accumulation of excessive watery fluid in the tissues.

Embryo. The product of conception during the period, roughly speaking, from fertilization to the time when the organism begins to assume the recognizable characteristics of its species, in man the organism from fertilized ovum to the end of the eighth week of development. To be distinguished from "fetus."

Endocrine glands. Those glands that pour their secretions directly into the bloodstream rather than through ducts. These glands are the pituitary, the thyroid, the parathyroids, the thymus, the pancreas, the adrenals, the ovaries in the female, and the testes in the male.

Enzyme. Any of various substances secreted by the body cells that act as expediters of the chemical processes necessary for the functioning of the organism.

Esophagus. The gullet; the portion of the digestive tract between the throat and the stomach, situated behind the air tube (trachea).

Eustachian tube. The auditory tube that passes from the inside of the eardrum (tympanic membrane) and eardrum cavity to the nasal part of the throat just above the soft palate. It is through this pharyngeal opening that air and fluid as well as infective organisms pass through the tube into the middle ear structures.

Exophthalmos. Protrusion of the eyeballs.

Fetus. The organism from the beginning of the ninth week to birth.

Gamma globulin. The concentration of that portion of the blood serum containing the antibodies to the infections from which the individual has suffered in the past.

Gastrointestinal system. The digestive system, but particularly the stomach and intestines.

Gene. The physical unit of heredity, a small region in a chromosome, consisting of a giant molecule or part of such a molecule consisting mainly of deoxyribonucleic acid (DNA).

Genitourinary system. The system consisting of the kidneys, ureters, bladder, external genitalia, and their accompanying structures.

Gland. A secreting organ.

Hemolytic diseases of the newborn. Diseases of the blood of newborn infants, in which red blood cells are destroyed, resulting in the diffusion through the body of the breakdown products of the red pigment of the blood, the hemoglobin, which are particularly poisonous to brain cells.

Hormone. A chemical substance, formed in one organ or part of the body and carried in the blood to another organ or part which it stimulates to functional activity. More than fifty hormones have thus far been recognized, including the sex hormones.

Hydramnios. Excessive amniotic fluid in the amniotic sac, which may produce considerable enlargement of the mother's abdomen.

Hyperthyroidism. Excessive secretion of thyroid hormone. Occurs eight times as frequently in women as in men. The principal symptoms are protruding eyeballs, tremor of the hands, rapid heartbeat with palpitations, fatigue, loss of weight, and thyroid enlargement. Several satisfactory forms of treatment exist.

Hypoglycemia. An abnormally small concentration of sugar in the blood.

Hypothalamus. An area at the base of the brain situated above the pituitary gland, the hinder portion of which forms part of this structure, together with the stalk of the pituitary, and a variety of other concentrations of nerve cells or nuclei.

Hypothyroidism. Often termed myxedema. Undersecretion of the thyroid hormone. Congenital hypothyroidism due to severe thyroid deficiency produces cretinism. Hypothyroidism usually occurs in adults. The disorder has a predilection for women who have had many children. In the severe forms of the disorder, symptoms are lethargy, somnolence, marked slowing of speech and intellectual processes, and a failing memory. Nose and lips tend to become thick, the eyelids puffy. The tongue is enlarged, the hair of the head is lusterless and falls out. This is also true of the eyebrows. There are many other symptoms. Treatment is most satisfactory with dried thyroid gland taken by mouth. The hypothyroidism of pregnancy is not usually of the severe kind.

Leukocyte. Any one of the white blood cells.

Luteotrophin. A hormone of the anterior part of the pituitary gland, which acts to maintain the corpus luteum of the ovary and promotes the formation of progesterone.

Malformation. Any deformity, either congenital or acquired as a result of disease, disorder, or injury.

Malocclusion. Any deviation from the normal areas of contact between the teeth of the upper and lower jaws.

Maturity. In the particular context of this book the term is used to refer to the period at which the female organism becomes capable of optimally maintaining a pregnancy. This period commences, on the average, at twenty-three years, give or take two years, and lasts approximately five years.

Meconium. The material in the intestinal tract of the fetus and the first discharges therefrom in the newborn infant, consisting of epithelial cells, mucus, and bile. It is of a greenish color.

Metabolism. The chemical and physical changes continuously going on in living organisms and cells, whereby food is assimilated and built up (anabolism), and broken down (catabolism), thus effecting the nutrition of the organism.

Metabolism, basic rate of. The heat production of an individual at the lowest level of cell chemistry in the waking state. It is determined in the individual before breakfast, at complete bodily and mental rest at a room temperature of 68°F.

Miscarriage. Loss of the product of conception prior to the twentieth week.

Mongolism. *See* Down's syndrome.

Multipara. A woman who has had two or more pregnancies.

Neurochemical. Pertaining to the chemical changes that result from the functioning of the nervous system or to the effect of the interaction between nervous and chemical processes.

Nubility. The period of development during which the functions of ovulation and reproductive potentiality are established in the female.

Organogenesis. The formation of organs during development. The period of organ formation in the human embryo begins at the close of the fourth week, and the basic patterns of the organs are laid down by the end of the seventh week.

Ovum. The female egg or sexual cell, from which, by union with the male sexual cell or spermatozoon, a new individual is developed.

Oxytocic drug. A drug that hastens childbirth or promotes the rapidity of labor.

Pancreas. The abdominal gland that secretes pancreatic juice, which is discharged into the intestine, and the hormone insulin.

Phocomelia. Abnormality of development in which the individual's hands and feet seem to be attached directly to the shoulders and hips without, or with only slight, intervention of arms and legs.

Placenta. The organ of communication between the fetus and the mother. The umbilical cord proceeds from the smooth fetal side of the placenta; on the maternal side there are a number of lobules or cotyledons, which project into spaces (sinuses) in the wall of the uterus, and are there bathed in maternal blood; in this manner the fetal blood receives oxygen and nutriment and yields waste material. The placenta, therefore, performs the functions of respiration, nutrition, and excretion. After expulsion of the child, the placenta is called the afterbirth.

Placenta previa. The condition in which the placenta is implanted in the lower segment of the uterus, extending to the margin of the internal opening of the neck of the uterus (the cervix) or partially or completely obstructing the opening.

Plasma. The fluid portion of the blood.

Prediabetes. Predisposition to diabetes.

Pre-eclampsia. The nonconvulsive stage of an acute hypertensive disorder of pregnancy. See Toxemia of pregnancy.

Prematurity. Referring to the condition of being born in an immature state denoted by a birth weight of not more than 5 pounds 8 ounces.

Presentation of fetus. The part of the body of the fetus which is in advance during birth. The back of the head (occiput), chin, and sacrum (lower back) are the determining points in vertex, face, and breech presentations.

Primigravida. A woman who is pregnant for the first time.

Primipara. A woman who has delivered only one viable child.

Progesterone. The hormone of the corpus luteum of the ovary and placenta which produces changes favoring the maintenance of pregnancy. These changes are in the lining wall of the uterus (endometrium), adapting it for reception of the fertilized ovum, inhibition of the uterine musculature, stimulation of the mammary glands, relaxation of the pelvic ligaments, and the like.

Protozoon (plural: protozoa). A one-celled organism belonging to the animal kingdom. Some protozoa are parasitic in animals and man, causing such diseases as malaria, toxoplasmosis, sleeping sickness, and amebic dysentery.

Puberty. The period of inception of those changes in the reproductive system that are exhibited in the gradual development of the secondary sexual characteristics and the appearance of the first menstruation (menarche). The first vaginal bleeding is not usually coincident with ovulation and therefore with the capacity to conceive or reproduce. There is an adolescent sterility interval from the first menstruation to the capacity to reproduce which, on the average, lasts about three years (See Ashley Montagu, *The Reproductive Development of the Female*. New York: Julian Press, 1957).

Rh factor. A complex antigenic substance occurring on the surface of red blood cells. When a mother who is Rh-negative (that is, one whose blood lacks this substance) carries an Rh-positive child, subsequent Rh-positive children may be born with hemolytic diseases of the newborn.

Sensitization. The production by the body cells of a specific ferment which acts upon the protein whose presence has called it into existence; when the organism reacts to the special protein, it is said to be sensitized to that protein.

Sperm or Spermatozoon. The male sexual cell.

Spleen. The large organ lying in the upper part of the abdominal cavity on the left side, between the stomach and the diaphragm. The spleen is one of the organs involved in blood formation.

Stillbirth. Fetus born dead after twenty-eight completed weeks of gestation.

Tetany. A disorder marked by intermittent, painful muscular contractions, local losses of sensation, gastric and intestinal difficulties, and general nervous irritability.

Thalamus. The large egglike mass of gray substance, with flattened inner surface, in the middle of the brain, which serves as a sorting station for incoming stimuli to the gray matter (cortex) of the brain. It also has important integrative and nonspecific functions, and also is secretory, directly affecting the pituitary gland through its stalk below the thalamic region.

Thyroid gland. The large endocrine gland in the neck which secretes the active iodine compound thyroxin.

Toxemia of pregnancy. A condition of the pregnant female characterized by rapid gain of weight, elevation of blood pressure, and appearance of albumin in the urine. In most cases it can be prevented by competent medical care during the antenatal period.

Urogenital system. The system comprising the kidneys, ureters, bladder, and internal and external genital organs.

Vasoconstrictor drug. A drug having the effect of narrowing the diameters of the blood vessels.

Villi (singular: villus). A minute projection from the surface, usually of a mucous membrane, such as those in the intestines. The fetal villi, arising from the covering membrane of the amniotic sac (the chorion), penetrate the maternal uterine wall and come to rest in the uterine blood spaces (sinuses). There are about 800,000 placental villi. Laid end to end, these villi would extend for over 11 miles, and their surface area would cover about 140 square feet.

Virus. A chemical entity on the borderline between the living and the nonliving; viruses cause disease in plants, bacteria, insects, animals, and man.

Vulva. The external genitalia of the female, including the labia, the clitoris, and the entrance to the vagina.

INDEX